Lectures in Mathematics
ETH Zürich
Department of Mathematics
Research Institute of Mathematics

Mariano Giaquinta

Introduction to Regularity Theory for Nonlinear Elliptic Systems

1993

Birkhäuser Verlag
Basel · Boston · Berlin

Author:

Mariano Giaquinta
Dipartimento di Matematica Applicata
Università di Firenze
Via S. Marta 3
I-50139 Firenze
Italia

A CIP catalogue record for this book is available from the Library of Congress, Washington D.C., USA

Deutsche Bibliothek Cataloging-in-Publication Data
Giaquinta, Mariano:
Introduction to regularity theory for nonlinear elliptic systems / Mariano Giaquinta. – Basel ; Boston ; Berlin : Birkhäuser, 1993
(Lectures in mathematics)
ISBN 3-7643-2879-7 (Basel ...)
ISBN 0-8176-2879-7 (Boston)

Printed on acid-free paper produced of chlorine-free pulp
Printed in Germany
ISBN 3-7643-2879-7
ISBN 0-8176-2879-7

9 8 7 6 5 4 3 2 1

Contents

Preface

These notes are the result of a course I have given during the winter semester 1983/84 for the Nachdiplomstudium in Mathematik at the Mathematikdepartement of ETH, Zürich, where I was invited by Professor Jürgen Moser. It is a pleasure for me to thank him for this invitation and for his constant interest.

These notes have been written by Enrico Leuzinger, and I would like to thank him for his careful work.

I also would like to acknowledge with gratitude the hospitality of the Mathematkikdepartement and the support of the Forschungsinstitut für Mathematik of ETH.

The lectures are by no means a survey, but only an introduction, among the many that could be given, to the regularity theory of minimizers of regular integrals in the Calculus of Variations and of solutions of nonlinear elliptic systems. For instance I have not presented many important topics and I have always restricted myself to simple situations avoiding technical (and sometimes even nontechnical) difficulties. The reader who wants to have more information on the topics treated can refer for example to Bers-Schechter [5], Morrey [55], Ladyzhenskaya-Ural'tseva [49], Gilbarg-Trudinger [35], and Giaquinta [22]; this last reference being the most related one.

In chapter 1 we describe quickly the Hilbert-space approach to the existence of solutions to nonlinear systems in divergence form and direct methods in the Calculus of Variations.

In chapter 2 we deal with the so-called Hilbert-space approach to the regualrity.

In chapter 3 we give a proof of Schauder-estimates, which is free from potential theory, following some ideas of Morrey [54] and Campanato [8] and the presentation of Campanato [8], see also Giusti [38].

Also L^p-theory, chapter 4, is developped free from potential theory on the basis of an interpolation theorem by Stampacchia [69], see also Campanato-Stampacchia [10], and does *not* use the well-known theorem of Calderon-Zygmund on singular integrals.

In chapter 5 we deal with De Giorgi's class and De Giorgi's theorem [11], which play a fundamental role in the theory of regularity for scalar equations, as well as with Moser's iteration technique [58] which leads to Harnack's inequality. Then by combining De Giorgi's results together with some ideas of Krylov-Safanov [48], we prove, following a recent work by Di Benedetto-Trudinger [13], that Harnack's inequality also holds for functions in the De Giorgi class. Finally we hint at the notion of quasiminima [26], [30], which plays a relevant, unifying and clarifying role in the regularity theory, and we prove a general regularity theorem for minimizers in the scalar case due to Giaquinta-Giusti [26], [30].

Finally chapter 6 deals with the regularity theory in the vector valued case, but we shall only describe a few model results, refering the reader to Giaquinta [22] for more information. We finish with some theorems concerning the regularity of harmonic mappings between Riemannian manifolds.

Mariano Giaquinta

Chapter 1

Direct methods in the calculus of variations: existence

In 1900 D. Hilbert stated as *20th problem:* "Has not every regular variational problem a solution, provided certain assumptions regarding the given boundary conditions are satisfied, and provided also if necessary that the notion of a solution shall be suitably extended?"

The aim of this chapter is twofold: first to provide a positive answer to Hilbert's problem in some simple situations and second to illustrate how "extending suitably the notion of a solution" the existence problem simplifies a great deal and actually becomes solvable.

In the first section we illustrate the so-called "Hilbert-space-approach" to boundary value problems (mainly of Dirichlet-type) for linear elliptic systems, which originated from the classical "Dirichlet-principle" and which provides a good understanding of it.

In section two we describe and use in very simple situations the so-called "Direct methods in the calculus of variations". In this way we state a few basic existence results both for linear boundary value problems and minimum problems for variational integrals and we will see the key role of the notion of "ellipticity".

1.1 Hilbert-space approach: Existence theorems

We first recall
Riesz's representation-theorem: *Let* $(H, (\ ,\ ))$ *be a Hilbert-space and* F *a*

bounded linear functional on H. Then there exists a unique $u_F \in H$ such that $F(v) = (u_F, v)$ for $\forall v \in H$ and moreover $||u_f|| = ||F||$.

Let us now look at an open, bounded and connected set $\Omega \subset \mathbb{R}^n$ with a smooth boundary $\partial\Omega$.

We define:
$H^{1,2}(\Omega) := \Big\{u \in L^2(\Omega) \mid u$ has first derivatives in the sense of distributions in $L^2(\Omega)\Big\}$

$$||u||_{H^{1,2}(\Omega)} := \left(\int_\Omega |u|^2 dx + \int_\Omega |Du|^2 dx\right)^{1/2}$$

$H_0^{1,2}(\Omega) := \overline{C_0^1(\Omega)}$ where the completion is to understand with respect to the norm $||.||_{H^{1,2}(\Omega)}$. Clearly $H_0^{1,2}(\Omega) \subset H^{1,2}(\Omega)$.

Observe also that in $H_0^{1,2}(\Omega)$ the norm $\left(\int |Du|^2 dx\right)^{1/2}$ is equivalent to $||u||_{H^{1,2}(\Omega)}$. That follows from

Poincaré-inequality: *If $u \in H_0^{1,2}(\Omega)$ we have $\int_\Omega |u|^2 dx \le c(\Omega) \int_\Omega |Du|^2 dx$.*

PROOF: Assume Ω can be enclosed in a cube $Q = \{x \in \mathbb{R}^n \mid |x_i| \le a\,,\ 1 \le i \le n\}$ and $D = D_{x_1}$ then for any $x \in Q$ we have

$$u^2(x) = \left(\int_{-a}^{x_1} Du(t, x_2, \ldots, x_n) dt\right)^2 \le (x_1 + a) \int_{-a}^{x_1} (Du)^2 dt \le 2a \int_{-a}^{a} (Du)^2 dt$$

thus

$$\int_{-a}^{a} u^2(x_1, \ldots, x_n) dx_1 \le 4a^2 \int_{-a}^{a} (Du)^2 dx_1 \quad .$$

Integration over $x_2, \ldots, x_n$ from $-a$ to a gives the result.

To get some other Poincaré-inequalities we recall

Rellich's theorem: *Let Ω be a bounded smooth domain in $\mathbb{R}^n$ (say with locally continuous boundary), then the immersion*

$$H^{1,2}(\Omega) \hookrightarrow L^2(\Omega)$$

is a compact operator.

For a proof cf. [60].

Now suppose that we can find $u_k \in H_0^{1,2}(\Omega)$ for each $k \in \mathbb{N}$ such that

(i) $\int_\Omega |u_k|^2 dx \ge k \int_\Omega |Du_k|^2 dx$

(ii) $\int_\Omega u_k dx = 0$

(iii) $\int_\Omega |u_k|^2 dx = 1$.

By Rellich's theorem there exists a subsequence $u_{k_j} \xrightarrow{L^2(\Omega)} u$ and $Du_{k_j} \rightharpoonup Du$ but from (i) we see that then $Du = 0$ and hence $u = 0$ (because of (ii)), that is a contradiction to (iii).

This argument proves the following:

If in Ω Rellich's theorem holds, i.e. the immersion $H^{1,2}(\Omega) \hookrightarrow L^2(\Omega$ is compact, then there exists a constant c depending on Ω such that

$$\int_\Omega |u - u_\Omega|^2 dx \le c \int_\Omega |D(u - u_\Omega)|^2 dx = c \int_\Omega |Du|^2 dx \ ,$$

where $u_\Omega \equiv \fint_\Omega u dx := \frac{1}{|\Omega|} \int_\Omega u dx$ and $|\Omega| := \int_\Omega dx$.

In case $\Omega = B_R :=$ Ball of Radius R or $\Omega = B_R \backslash B_{R/2}$ the following *rescaling argument* shows the explicit dependence of $c(\Omega)$ on R: Set first $R = 1$ to get a c such that

$$\int_{B_1} |u - u_{B_1}|^2 dx \le c \int_{B_1} |Du|^2 dx$$

then for arbitrary R:

$$\begin{aligned} \int_{B_R} |u(y) - u_{B_r}|^2 dy &= \int_{B_1} |u(Rx) - u_{B_R}|^2 R dx \\ &= R \int_{B_1} |\tilde{u}(x) - \tilde{u}_{B_1}|^2 dx \le Rc \int_{B_1} |D\tilde{u}|^2 dx \\ &= R^3 c \int_{B_1} |Du(Rx)|^2 dx \\ &= cR^2 \int_{B_R} |Du(y)|^2 dy \end{aligned}$$

i.e. we have the

Poincaré-inequalities:

$$\int_{B_R} |u - u_{B_r}|^2 dx \le c(n) R^2 \int_{B_R} |Du|^2 dx$$

and also

$$\int_{B_r \backslash B_{R/2}} \left| u - u_{B_r \backslash B_{R/2}} \right|^2 dx \le c(n) R^2 \int_{B_r \backslash B_{R/2}} |Du|^2 dx \ .$$

$H^{1,2}(\Omega)$ and $H_0^{1,2}(\Omega)$ are Hilbert-spaces with inner product respectively

$$\begin{aligned}(u,v) &= \int_\Omega uv\,dx + \int_\Omega DuDv\,dx \\ (u,v) &= \int_\Omega DuDv\,dx \ .\end{aligned}$$

Let us consider the linear functional

$$\begin{aligned}F : H^{1,2}(\Omega) &\longrightarrow \mathbb{R} \\ u &\longmapsto \int_\Omega fu\,dx + \int_\Omega \sum_\alpha f_\alpha D_\alpha u\,dx\end{aligned}$$

where $f, f_\alpha \in L^2(\Omega)$.

F is a bounded linear functional on $H^{1,2}(\Omega)$ (or on $H_0^{1,2}(\Omega)$). Thus from Riesz's theorem in $H_0^{1,2}(\Omega)$ *we get a unique* $u \in H_0^{1,2}(\Omega)$, *such that*

$$(*) \qquad \int_\Omega DuDv\,dx = \int_\Omega fv\,dx + \int_\Omega \sum_\alpha f_\alpha D_\alpha v\,dx \text{ for } \forall v \in H_0^{1,2}(\Omega) \ .$$

We express this by saying, that u *is a weak solution (or a solution in the sense of distributions) of the Dirichlet boundary-value-problem:*

$$(**) \qquad \begin{cases} -\Delta u = f - D_\alpha f_\alpha & \text{in } \Omega \\ u = 0 & \text{on } \partial\Omega \end{cases}$$

Note that if u is a classical solution of $(**)$we can deduce $(*)$ by multiplying the differential equation of $(**)$by a $\phi \in C_0^\infty(\Omega)$ and integrating by parts; while from $(*)$ if u is sufficiently smooth we derive $(**)$.

Let us now look at

$$a(u,v) = \int_\Omega A^{\alpha\beta}(x) D_\alpha u D_\beta v\,dx$$

with

1) $A^{\alpha\beta}(x) \in L^\infty(\Omega)$
2) $A^{\alpha\beta} = A^{\beta\alpha}$
3) $A^{\alpha\beta}(x)\xi_\alpha\xi_\beta \geq \nu|\xi|^2$, $\nu > 0$, for $\forall \xi \in \mathbb{R}^n$.

Condition 3) is referred as *ellipticity.* $a(u,v)$ defines a bounded bilinear form on $H_0^{1,2}(\Omega)$. Because of ellipticity $a(u,v)$ is *coercive,* i.e. $a(u,u) \geq \nu \int_\Omega |Du|^2 dx$ hence $a(u,v)$ is an inner product on $H_0^{1,2}(\Omega)$ (equivalent to the given one).

Again we can apply Riesz's representation theorem to the functional F (defined on p. 4): *there exists a unique $u \in H_0^{1,2}(\Omega)$ such that*

$$(+) \qquad \int_\Omega A^{\alpha\beta} D_\alpha u D_\beta v dx = \int_\Omega \left[f_0 v + \int_\alpha f_\alpha D_\alpha v \right] dx \ .$$

Integration by parts leads to the weak solution of

$$\begin{cases} -D_\alpha(A^{\alpha\beta}(x) D_\beta u) &= g_0 + \sum_\alpha f_\alpha D_\alpha \quad \text{in} \quad \Omega \\ u &= 0 \qquad\qquad\qquad \text{on} \quad \partial\Omega \ . \end{cases}$$

Exercise: Prove that the unique point u which is a solution of (+) is the unique minimum point of the functional

$$\mathbb{F}(u) = \frac{1}{2} \int_\Omega A^{\alpha\beta} D_\alpha u D_\beta u dx + \int_\Omega \sum_\alpha f_\alpha D_\alpha u dx \ .$$

Remark: Let us look at the non-homogeneous Dirichlet boundary value problem

$$\begin{cases} -D_\alpha(A^{\alpha\beta}(x) D_\beta u) &= 0 \quad \text{in} \quad \Omega \\ u &= u_0 \quad \text{on} \quad \partial\Omega \end{cases}$$

Suppose that there exists $\bar{u} \in H^{1,2}(\Omega)$ such that $\bar{u}\Big|_{\partial\Omega} = u_0$ then we can reduce the above problem to a homogeneous one: just set $w := u - \bar{u}$ to get:

$$\begin{cases} -D_\alpha(A^{\alpha\beta}(x) D_\beta (u - \bar{u}) &= -D_\alpha \left(A^{\alpha\beta}(x) D_\beta \bar{u} \right) \\ w &= u - \bar{u} \in H_0^{1,2}(\Omega) \end{cases}$$

so $w + \bar{u}$ is a solution of the nonhomogeneous problem.

In general there does not exist such an $\bar{u}$: One can show that the linear operator trace:

$$\begin{array}{ccc} C_0^1(\bar{\Omega}) & \longrightarrow & C^0(\partial\Omega) \\ u \longmapsto & u\Big|_{\partial\Omega} & \end{array}$$

can be extended continuously to a linear operator trace: $H^{1,2}(\Omega) \to L^2(\partial\Omega)$ but in general we have trace $\left(H^{1,2}(\Omega)\right) \subsetneq L^2(\partial\Omega)$ and $C^0(\partial\Omega) \not\subset \text{trace}\left(H^{1,2}(\Omega)\right)$.

With the same assumptions 1), 2, 3) on $A^{\alpha\beta}$ as on p. 4 $b(u,v) = a(u,v) + \lambda^2 \int_\Omega uv dx$ is an inner product on $H^{1,2}(\Omega)$. So by Riesz's theorem there exists a unique $u \in H^{1,2}(\Omega)$ such that

$$(++) \quad \int_\Omega A^{\alpha\beta} D_\alpha u D_\beta v dx = \int_\Omega \left[f_0 + \sum f_\alpha D_\alpha - \lambda^2 u \right] v dx \text{ for } \forall v \in H^{1,2}(\Omega) \ .$$

In particular (++) holds for all $v \in H_0^1(\Omega)$ and thus

$$-D_\beta(A^{\alpha\beta}D_\alpha u) = f_0 - D_\alpha f_\alpha - \lambda^2 u \ .$$

On the other hand the left hand side of (++) is equal to

$$\begin{aligned}&\int_\Omega D_\beta(A^{\alpha\beta}D_\alpha uv)dx - \int_\Omega D_\beta(A^{\alpha\beta}D_\alpha u)vdx\\ &\qquad = \int_{\partial\Omega} A^{\alpha\beta}D_\alpha u\nu_\beta v d\sigma - \int_\Omega D_\beta(A^{\alpha\beta}D_\alpha u)vdx\end{aligned}$$

hence

$$\int_{\partial\Omega} A^{\alpha\beta}D_\alpha u\nu_\beta v d\sigma = 0 \ .$$

u thus solves the *Neumann-boundary value problem:*

$$\begin{cases} -D_\alpha\left(A^{\alpha\beta}(x)D_\beta u\right) + \lambda^2 &= f_0 + \sum\limits_\alpha D_\alpha f_\alpha \\ A^{\alpha\beta}D_\beta u\cdot\nu_\alpha &= 0\end{cases}$$

$\frac{du}{d\nu^*} = A^{\alpha\beta}D_\beta u\cdot\nu_\alpha$ is called the *conormal derivative.*

Example: For $A^{\alpha\beta} := \delta^{\alpha\beta}$ we find

$$\begin{cases} -\Delta u + \lambda^2 u &= f_0 + \sum\limits_\alpha D_\alpha f_\alpha \\ \frac{du}{d\nu} &= 0 \ .\end{cases}$$

Similar considerations clearly hold for *systems* with the ellipticity-condition:

$$A_{ij}^{\alpha\beta}(x)\xi_\alpha^i\xi_\beta^j \geq \nu|\xi|^2 \quad \text{for} \quad \forall\xi\in\mathbb{R}^n$$

i.e., *for* $f^i, f_\alpha^i \in L^2(\Omega)$ *there exists a unique* $u \in H_0^{1,2}(\Omega,\mathbb{R}^N)$ *such that*

$$\int_\Omega A_{ij}^{\alpha\beta}(x)D_\alpha u^i D_\beta\phi^j dx = \int_\Omega f^i\phi^i dx + \int_\Omega f_\alpha^i D_\alpha\phi^i dx \quad \text{for} \quad \forall\phi\in H_0^1(\Omega,\mathbb{R}^N) \ .$$

Riesz's theorem requires symmetry.

More generally we have the

Theorem of Lax-Milgram *Let* $(H,(\ ,\ ))$ *be a Hilbert-space and* $a(u,v)$ *a bilinear form on* H *that is bounded,* i.e.

$$|a(u,v)| \leq c|u|\ |v| \quad \text{for} \quad \forall u,v\in H \ .$$

Suppose moreover that $a(u,v)$ is coercive, i.e.

$$A(u,u) \geq \nu|u|^2 \quad \text{for} \quad \forall u \in H, \nu > 0 \ .$$

Then for every bounded linear functional F on H there exists a unique u_F, such that $F(v) = a(u_F, v)$ for $\forall v \in H$.

Moreover we have $|u_F| \leq c\,||F||$.

PROOF: $a(u,v) = (Tu, v) = (u, T^*v)$ where T is a bounded linear operator.

Set $\tilde{a}(u,v) := (TT^*u, v) = (T^*u, T^*v)$. $\tilde{a}$ is bilinear, bounded, symmetric and coercive:

$$\nu|u|^2 \leq a(u,u) = (u, T^*u) \leq |u|\ |T^*u| = |u|(\tilde{a}(u,u))^{1/2} \ .$$

Hence $\tilde{a}$ is an inner product and by Riesz's theorem there exists a unique u'_F with

$$F(v) = \tilde{a}(u'_F, v) = (TT^*u'_F, v) = a(T^*u'_F, v) \quad \text{for} \quad \forall v \in H \ .$$

So $u_F = T^*u'_F$.

By this theorem the existence problem is reduced to coerciveness.

Consider now a general second order integro-differential form:

$$a(u,v) = \int_\Omega \left[A^{\alpha\beta}(x)D_\alpha u D_\beta v + b^\alpha(x) u D_\alpha v + c^\alpha(x) v D_\alpha u + d(x)uv\right] dx$$

Proposition 1.1 *Suppose* 1) *$A^{\alpha\beta}, b^\alpha, c^\alpha, d \in L^\infty(\Omega)$ and* 2) *$A^{\alpha\beta}\xi_\alpha\xi_\beta \geq \nu|\xi|^2$ for $\forall \xi \in \mathbb{R}^n$, i.e. ellipticity, then, for $u \in H^{1,2}(\Omega)$, we have*

(*) $\qquad a(u,u) \geq \lambda_1||u||^2_{H^{1,2}(\Omega)} - \lambda_0||u||^2_{L^2(\Omega)}$ *with* $\lambda_1 > 0$.

PROOF: We use $ab \leq \varepsilon a^2 + \frac{1}{\varepsilon}b^2$ for arbitrary $\varepsilon > 0$.[1]

$$a(u,u) \geq c_1 \int_\Omega |Du|^2 - c_2 \int_\Omega uDu - c_3 \int_\Omega |u|^2$$

and $-c_2 \int_\Omega uDu \geq -c_2\varepsilon \int_\Omega |Du|^2 - c_2\varepsilon^{-1}\int_\Omega |u|^2$ choose ε in such a way that $\lambda_1 - c_1 - c_2\varepsilon > 0$.

We refer to (*) by saying that $a(u,v)$ is *weakly coercive.*

Proposition 1.2 *Suppose* 1) $A^{\alpha\beta} \in L^\infty; b^\alpha, c^\alpha \in L^n(\Omega); d \in L^{n/2}(\Omega)$ *and* 2) $A^{\alpha\beta}\xi_\alpha\xi_\beta \geq \nu|\xi|^2$ *for $\forall \xi \in \mathbb{R}^n$, then $a(u,v)$ is weakly coercive.*

[1]This is sometimes called Young's inequality.

PROOF: (hints): Use proposition 1.1 and the fact, that for $f \in L^p$ there exists $f_1 \in L^\infty$ and $f_2 \in L^p$ such that $f = f_1 + f_2 \& ||f_2||_{L^p} < \varepsilon$.
Remark:

1) Ellipticity depends only on the symmetric part of $A^{\alpha\beta}$:

$$A^{\alpha\beta} = 1/2(A^{\alpha\beta} + A^{\beta\alpha}) + 1/2(A^{\alpha\beta} - A^{\beta\alpha}) \ .$$

2) Propositions 1.1 and 1.2 also hold for systems provided that

$$A_{ij}^{\alpha\beta}\xi_\alpha^i\xi_\beta^j \geq \nu|\xi|^2 \quad \text{for} \quad \forall\xi \in \mathbb{R}^{nN} \ .$$

We have seen that ellipticity implies weak coerciveness. In the scalar case (i.e. $N = 1$) we have also the opposite.

Theorem 1.1 *If $a(u,u) \geq c\int_\Omega |Du|^2 dx$ for $\forall u \in H_0^{1,2}(\Omega)$ then $A^{\alpha\beta}\xi_\alpha\xi_\beta \geq c|\xi|^2$ for $\forall\xi \in \mathbb{R}^n$.*

PROOF: We may assume Re $a(u,u) \geq c\int_\Omega |Du|^2 dx$ for all $u \in H_0^{1,2}(\Omega, \mathbb{C})$. Now set $u(x) := \Phi(x)e^{\tau i x\cdot\xi}$ with $\phi \in C_0^\infty(\Omega)$. We have $D_\alpha u = (D_\alpha\phi - \tau i\xi_\alpha\phi)e^{\tau i x\cdot\xi}$ and

$$\text{Re}\int_\Omega A^{\alpha\beta}D_\alpha u\overline{D_\beta u}dx = \text{Re}\int_\Omega \tau^2 A^{\alpha\beta}\xi_\alpha\xi_\beta\phi^2 dx + 0(\tau^2)$$

on the other hand

$$\int_\Omega |Du|^2 dx = \int_\Omega \tau^2|\xi|^2\phi^2 dx + 0(\tau^2) \ .$$

We devide by τ^2 and use the assumption to get

$$\text{Re}\int_\Omega A^{\alpha\beta}\xi_\alpha\xi_\beta\phi^2 dx \geq c\int_\Omega |\xi|^2\phi^2 dx + \frac{0(\tau^2)}{\tau^2}$$

hence taking the limit $\tau \to \infty$ and using Lebesgue convergence theorem we finish the proof.

Similar computations can be carried on for *systems:*

$$(*) \qquad -D_\alpha(A_{ij}^{\alpha\beta}D_\beta u^i) = 0 \qquad j = 1, \ldots, N$$

$a(u,v) = \int_\Omega A_{ij}^{\alpha\beta}D_\alpha u^i D_\beta v^j dx$ is positive definite if we assume

$$A_{ij}^{\alpha\beta}\xi_\alpha^i\xi_\beta^j \geq c|\xi|^2 \quad \text{for} \quad \forall\xi \in \mathbb{R}^{nN}$$

this is the so-called *Legendre-condition.*

If the $L - H$-condition holds we call $(**)$*strongly elliptic.*

Remark: Clearly the Legendre-condition implies the Legendre-Hadamard-condition (just take $\xi_\alpha^h := \xi_\alpha \eta^h!$).

But the converse is not true: $LH \not\Rightarrow L$.

Example: $n = N = 2$. Define $A_{ij}^{\alpha\beta}$ by

$$\sum_{i,j,\alpha,\beta=1}^{2} A_{ij}^{\alpha\beta} D_\alpha u^i D_\beta u^j := \det(D_\alpha u^i) + \varepsilon |Du|^2$$

LH-condition $\Leftrightarrow$

$$\begin{array}{rcll} A_{ij}^{\alpha\beta}\xi_\alpha\xi_\beta\eta^i\eta^j & = & \varepsilon|\xi|^2|\eta|^2 & \text{for } \forall \xi, \eta \in \mathbb{R}^2 \\ \Big(\det(\xi_\alpha \eta^i) & = & 0 & \text{for } n = N = 2!\Big) \end{array}$$

L-condition $\Leftrightarrow$

$$A_{ij}^{\alpha\beta}\xi_\alpha^i\xi_\beta^j = \det(\xi_\alpha^i) + \varepsilon|\xi|^2 \, .$$

Now for any $0 < \varepsilon < 1$ there exists a $\xi \in \mathbb{R}^2$ such that $\det(\xi_\alpha^i) + \varepsilon|\xi|^2 < 0$. We have seen that weak coerciveness implies a $L - H$-condition. Under additional assumptions the converse also holds. This is the content of

Gårding's inequality. *Suppose*

1) $A_{ij}^{\alpha\beta}(x)$ *is uniformly continuous*

2) $A_{ij}^{\alpha\beta}\xi_\alpha\xi_\beta\eta^i\eta^j \geq c|\xi|^2|\eta|^2$ *for* $\forall \xi, \eta \in \mathbb{R}^n$, i.e. *an* $L - H$*-condition*

3) b_{ij}^α, c_{ij}^α, $d_{ij} \in L^\infty(\Omega)$,
then, for $u, v \in H_0^{1,2}(\Omega)$,

$$\begin{aligned} a(u,v) = & \int_\Omega \Big[A_{ij}^{\alpha\beta}(x) D_\alpha u^i D_\beta v^j \\ & + b_{ij}^\alpha(x) u^i D_\alpha v^j + c_{ij}^\alpha(x) v^i D_\alpha u^j + d_{ij}(x) u^i v^j\Big]\, dx \end{aligned}$$

is weakly coercive: there exist $\lambda_0 > 0$ *and* λ_1 *such that*

$$a(u,u) \geq \lambda_0 \int_\Omega |Du|^2 dx - \lambda_1 \int_\Omega |u|^2 dx \, .$$

PROOF: We will proceed in three steps:

Step 1: Assume i) $A_{ij}^{\alpha\beta} = \text{constant}$, ii) $b_{ij}^{\alpha\beta} = c_{ij}^{\alpha\beta} = c_{ij}^{\alpha} = d_{ij} = 0$

Because $H_0^{1,2}(\Omega) = \overline{C_0^\infty(\Omega)}$ it is sufficient to prove the inequality for elements of $C_0^\infty(\Omega)$. The idea is to use Fourier-transformation

$$\hat{f}(x) := \int_\Omega f(y) e^{-2\pi i x \cdot y} dy$$

$$\widehat{D_\alpha f}(x) = -2\pi i x_\alpha \hat{f} \;, \quad ||\hat{f}||_{L^2} = ||f||_{L^2}$$

$$\begin{aligned}
a(u,u) &= A_{ij}^{\alpha\beta} \int_\Omega \widehat{D_\alpha u}^i \overline{\widehat{D_\beta u}^j} dx \\
&= (2\pi)^2 A_{ij}^{\alpha\beta} \int_\Omega x_\alpha x_\beta \hat{u}^i \overline{\hat{u}^j} dx \\
&\geq c \int_\Omega |x|^2 |\hat{u}|^2 dx \\
&= (2\pi)^2 \tilde{c} \delta_{ij} \delta^{\alpha\beta} \int_\Omega x_\alpha x_\beta \hat{u}^i \overline{\hat{u}^j} dx \\
&= \tilde{c} \delta_{ij} \delta^{\alpha\beta} \int_\Omega \widehat{D_\alpha u}^i \overline{\widehat{D_\beta u}^j} dx \\
&= \tilde{c} \int_\Omega \left| \widehat{D_\alpha u}^j \right|^2 dx \\
&= \tilde{c} \int_\Omega \left| D_\alpha u^j \right|^2 dx \;.
\end{aligned}$$

Step 2: Suppose $u \in C_0^\infty(\Omega)$ with supp $u \subset U_\varepsilon(x_0)$.

$$\begin{aligned}
\int_\Omega A_{ij}^{\alpha\beta}(x) D_\alpha u^i D_\beta u^j dx &= \int_\Omega A_{ij}^{\alpha\beta}(x_0) D_\alpha u^i D_\beta u^j dx + \\
&\quad + \int_\Omega \left(A_{ij}^{\alpha\beta}(x) - A_{ij}^{\alpha\beta}(x_0) \right) D_\alpha u^i D_\beta u^j dx \\
&\geq c \int_\Omega |Du|^2 dx - \omega(|x - x_0|) \int_\Omega |Du|^2 dx
\end{aligned}$$

(choose ε in such a way that $c - w > 0$!).

Step 3: Cover $\bar{\Omega}$ with finitely many balls $B_\varepsilon(x_j) x_j \in \Omega$, $j = 1, \ldots, N$. Now choose $\phi_j \in C_0^\infty(B_\varepsilon(x_j))$ such that $\sum_{j=1}^N \phi_j^2 = 1$. Moreover we now take $a(u,v)$ in its general form:

$$\begin{aligned}
a(u,v) &= \int_\Omega \left[A_{ij}^{\alpha\beta} \left(\sum_k \phi_k^2 \right) D_\alpha u^i D_\beta u^j + \ldots \right] dx \\
&= \sum_k \int_\Omega A_{ij}^{\alpha\beta} D_\alpha(\phi_k u^j) D_\beta(\Phi_k u^i) dx + \\
&\quad \text{(terms which do not contain products of derivatives of } u) \geq \\
&\geq c \sum_{k=1}^N \int_\Omega |D(\phi_k u)|^2 \, dx - d||u||_{L^2} ||Dz||_{L^2} - e||u||_{L^2}^2 \\
&\quad \underset{\text{step2}}{\uparrow} \\
&= c \sum_{k=1}^N \int_\Omega \left[\phi_k^2 |Du|^2 + (D\phi_k)^2 u^2 + 2u\phi_k Du D\phi_k \right] dx - \\
&\quad -D||u||_{L^2} ||Du||_{L^2} - e||u||_{L^2}^2 \\
&\geq c \int_\Omega |Du|^2 dx - \tilde{d}||u||_{L^2} ||Du||_{L^2} - \tilde{e}||u||_{L^2}^2
\end{aligned}$$

now by Young's inequality weak coerciveness of $a(u,v)$ follows.

Remarks:

Weak coerciveness implies the existence of solutions for the Dirichlet boundary value problem:

$$\begin{cases} -D_\alpha(A_{ij}^{\alpha\beta} D_\beta u^j) + \lambda u &= f_0 + f_i^\alpha D_\alpha \quad \text{on } \Omega \\ u &= 0 \quad \text{on } \partial\Omega \end{cases}$$

for $\lambda > \lambda_1$.

One can show that actually there exist solutions for all λ except for an at most countable set of values λ, which are eigenvalues.

For more informations on the theory of coerciveness we refer to e.g. [1], [5], [60].

1.2 Direct methods in the calculus of variations: semicontinuity-theorems

In the Hilbert-space approach to boundary-value problems for linear systems coerciveness was the essential point.

If we want to minimize a functional

$$\mathbb{F}(u;\Omega) = \int_{\Omega} F(x, Du)dx$$

in some set K (= space of admissible functions, e.g. the Sobolev-space $H^{1,m}(\Omega) = \{u \in L^m(\Omega) \mid Du \in L^m(\Omega)\}$) the main point will be semicontinuity-theorems (cf. the remark on p. 13).

We call $u \in K$ a *minimizer* or *minimum-point* iff $\mathbb{F}(u) \leq \mathbb{F}(v)$ for $\forall v \in K$. The idea is now to use the so-called *direct methods in the calculus of variations.* Roughly we may describe them as follows:

The set K of admissible functions is not a priori equipped with a topology. So, on the ground of the well known Weierstrass-Fréchet-theorem, the minimum-problem

$$\mathbb{F}(u;\Omega) \rightarrow \min \quad \text{in } K$$

can be seen as the problem of introducing a topology on K for which *K is sequentially compact and $\mathbb{F}$ is sequentially lower semicontinuous* (s.l.s.c.).

In order to have $\mathbb{F}$ s.l.s.c. we need in general a rich topology, while for the compactness of K the topology needs not to be too rich. But we shall see that this compromise can be reached for a large class of functionals $\mathbb{F}$ in Sobolev-spaces $H^{1,m}$. We will see that lower semicontinuity theorems with respect to the weak convergence in $H^{1,m}$ hold true and that the compactness of K can simply be obtained by requiring suitable growth conditions on F.

Remark: The existence of a minimizer u for $\mathbb{F}$ on K implies a $L-H$-condition for F at u: Suppose that for $t \in (-1,1)$ and for $\forall\phi \in C_0^\infty(\Omega) u + t\phi \in K$ and consider

$$t \longmapsto \mathbb{F}(u + t\phi; \Omega) \ .$$

If u is a minimizer in K, then we have

$$\left.\frac{d\mathbb{F}}{dt}\right|_{t=0} = 0$$

and

$$\left.\frac{d^2\mathbb{F}}{dt^2}\right|_{t=0} \geq 0 \ .$$

Therefore we get

$$\int_\Omega F_{p^i_\alpha p^j_\beta}(x,u,Du)D_\alpha\phi^i D_\beta\phi^j dx \geq 0$$

and (cf. section 1.1)

$$F_{p^i_\alpha p^j_\beta}\xi_\alpha\xi_\beta\eta_i\eta_j \geq 0 \quad \text{for} \quad \forall\xi,\eta\in\mathbb{R}^n \ .$$

Remark: *In the scalar case* ($N = 1$)

$$F_{p_\alpha p_\beta}\xi_\alpha\xi_\beta \geq 0 \Leftrightarrow F(x,u,p) \quad \textit{is convex in } p.$$

The first semicontinuity-theorem is due to Morrey (based on earlier results by Tonelli):

Semicontinuity-theorem. *Let us assume: For Ω bounded in $\mathbb{R}^n$:*

(i) $f(x,u,p) \geq 0$ *(or* $\geq \phi(x) \in L^1(\Omega)$*)*

(ii) F, F_p *are continuous*

(iii) F *is convex with respect to* p.

Then $\mathbb{F}(u;\Omega)$ is weakly lower semicontinuous in $H^{1,m}_{\text{loc}}(\Omega;\mathbb{R}^N)$ i.e. if u_k converge weakly in $H^{1,m}(D)$ to u for $\forall D \subset\subset \Omega$ then

$$\mathbb{F}(u;\Omega) \leq \liminf_{k\to\infty}\mathbb{F}(u_k;\Omega) \ .$$

Before proving the theorem, let us see how it implies *existence:* Suppose that

$$F(x,u,p) \geq c|p|^m, \quad m > 1 \ , \quad 0 < c = \text{const.}$$

and that for some $\phi \in H^{1,m}(\Omega,\mathbb{R}^N)$

$$\mathbb{F}(\phi;\Omega) < \infty \ .$$

Consider the minimum problem

$$\mathbb{F}(u;\Omega) \to \min; u - \phi \in H^{1,m}_0(\Omega,\mathbb{R}^N) \ .$$

Let $\{u_n\}$ be a minimizing sequence, i.e.

$$\lim\mathbb{F}(u_n;\Omega) = \inf\left\{\mathbb{F}(u;\Omega) \mid u - \phi \in H^{1,m}_0(\Omega,\mathbb{R}^N)\right\} =: \mu$$

then

$$c\int_\Omega |Du_n|^2 dx \leq \mathbb{F}(u_n;\Omega) \leq \text{const.} \quad \text{(independent of } n\text{)}$$

on the other hand, from Poincaré inequality, we obtain

$$\begin{aligned}
\int_\Omega |u_n|^m dx &\le \text{const}\left[\int_\Omega |u_n - \Phi|^m dx + \int_\Omega |\Phi|^m dx\right] \\
&\le \text{const}\int_\Omega |D(u_n - \Phi)|^m dx + \text{const}\int_\Omega |\Phi|^m dx \\
&\le \text{const} \quad (\text{independent of } n)\,.
\end{aligned}$$

In conclusion:

$$||u_n||_{H^{1,m}(\Omega,\mathbb{R}^N)} \le \text{const} \quad (\text{independent of } n)\,.$$

So because $H^{1,m}(\Omega)$ is reflexive ($m > 1$!) there exists a subsequence $u_{k_j} \rightharpoonup u$ in $H^{1,m}(\Omega)$. Now we apply the theorem and get

$$\mathbb{F}(u) \le \liminf_{k\to\infty} \mathbb{F}(u_k) = \lim_{k\to\infty} \mathbb{F}(u_k) = \mu$$

i.e. u is a minimum point.

Proof of the theorem: It is sufficient to consider the case $m = 1$ (if $u_k \overset{H^{1,m}}{\rightharpoonup} u$ then $u_k \overset{H^{1,1}}{\rightharpoonup} u$). Fix $D \subset\subset \Omega$ (i.e. D strictly contained in Ω) and suppose ∂D smooth. From Rellich's theorem follows that $u_k \xrightarrow{L^1{}_{\text{loc}}(\Omega,\mathbb{R}^N)} u$ and a.e. (eventually this holds only for a subsequence) because every weakly bounded sequence is bounded.

Take $\varepsilon > 0$, then there exists a compact subset $K \subset D$ such that we have

(1) **Egorov's theorem:**

$u_k \to u$ *uniformly on* K *and* $\left|\int_{\Omega\setminus K} dx\right| < \varepsilon$

(2) **Lusin's theorem:**
u *and* Du *are continuous on* K.

(3) **Lebesgues's absolute continuity theorem:**

$$\int_K F(x,u,Du)dx \ge \int_D F(x,u,Du)dx - \varepsilon\ .$$

As F is convex in p we can write

$$\begin{aligned}
&\int_K F(x,u_k,Du_k)dx \\
&\geq \int_K F_{p^i_\alpha}(x,u_k,Du)(D_\alpha u^i_k - D_\alpha u^i)dx + \int_K F(x,u_k,Du)dx \\
&= \int_K F(x,u_k,Du)dx + \int_K \underbrace{F_{p^i_\alpha}(x,u,Du)}_{\text{bounded}}\underbrace{\left(D_\alpha u^i_k - D_\alpha u^i\right)}_{\to 0}dx + \\
&+ \int_K \underbrace{F_{p^i_\alpha}(x,u_k,Du) - F_{p^i_\alpha}(x,u^i_k,Du)}_{\longrightarrow 0}\underbrace{\left(D_\alpha u^i_k - D_\alpha u^i\right)}_{\text{equibounded}}dx
\end{aligned}$$

So by taking the limit $k \to \infty$ we find:

$$\liminf_{k\to\infty} \int_K F(x,u_k,Du_k)dx \geq \int_K F(x,u,Du)dx \geq \int_D F(u,u,Du)dx - \varepsilon \ .$$

As this holds for all D and all $\varepsilon > 0$ we have proved

$$\liminf_{k\to\infty} \int_\Omega F(x,u_k,Du_k)dx \geq \int_\Omega F(x,u,Du)dx \ .$$

For Egorov's, Lusin's and Lebesgue's theorem compare [65].

Remarks:

1) The theorem is also true if (ii) is replaced by:
   F is a *Caratheodory-function* (i.e.
   F is measurable in x and continuous in (u,p)).

2) For $N = 1$ semicontinuity implies convexity. But for $N > 1$ semicontinuity implies only quasi convexity (cf. the next theorem due to Morrey).

We define semicontinuity with respect to Lipschitz-convergence:
u_k converges to u in the sense of *Lipschitz-convergence* iff

1) $u_k \to u$ uniformly

2) the Lipschitz-norms of u and u_k are equibounded.

Then we have the

Theorem. *If $\mathbb{F}$ is lower semicontinuous with respect to Lipschitz-convergence then for $\forall\Omega$, $\forall x_0 \in \mathbb{R}^n$, $\forall u_0 \in \mathbb{R}^N$, $\forall p_0 \in \mathbb{R}^{nN}$ and all $\phi \in C_0^1(\Omega, \mathbb{R}^N)$ we have*

$$\int_\Omega F(x_0, u_0, p_0 + D\phi)dx \geq \int_\Omega F(x_0, u_0, p_0)dx$$

$$\left(\text{or } \fint_\Omega F(x_0, u_0, p_0 + D\phi)dx \geq F(x_0, u_0, p_0)\right)$$

and such an F is called quasiconvex.

Remark: $\tilde{u}(x) = u_0 + p_0 \cdot x$ is a linear function. Define

$$\dot{\mathbb{F}}(u; \Omega) := \int_\Omega F(x_0, u_0, Du)dx$$

We have $D\tilde{u} = p_0$ hence

$$\begin{aligned} \dot{\mathbb{F}}(\tilde{u}, \Omega) &= \int_\Omega F(x_0, u_0, p_0)dx \leq \int_\Omega F(x_0, u_0, p_0 + D\phi)dx \\ &= \dot{\mathbb{F}}(u_0 + \phi; \Omega) \quad \text{for all } \phi \in C_0^\infty(\Omega, \mathbb{R}^N) \quad . \end{aligned}$$

Consequently we see that linear functions are minimizers for $\dot{\mathbb{F}}$ and quasiconvex functions can be characterized by this property.

PROOF OF THE THEOREM: For the sake of simplicity let us consider $F = F(p)$. We may assume that Ω is a cube (which we normalize to the unit-cube). A function $\xi \in C_0^\infty(Q)$ extends to a periodic function in $\mathbb{R}^n$. Define $\xi_\nu(x) := \nu^{-1}\xi(\nu x)\nu \in \mathbb{N}$ (where $\nu \to \infty$).

$$\begin{aligned} u_0(x) &:= u_0 + p_0 \cdot x \\ u_\nu(x) &:= u_0 + \xi_\nu(x) \ . \end{aligned}$$

Then $u_\nu(x) \to u_0(x)$ with respect to Lipschitz-convergence. So by assumption

$$|Q|F(p_0) \leq \liminf_{\nu\to\infty} \int_Q F(p_0 + D\xi_\nu(x))dx \ .$$

Now $D\xi_\nu(x) = (D\xi)(\nu x)$ and if we change variables, $\nu x = y$, we get

$$\begin{aligned} |Q|F(p_0) &\leq \liminf_{\nu\to\infty} \int_{\nu Q} F(p_0 + D\xi(y))dy \cdot \frac{1}{\nu^n} \\ &= \liminf_{\nu\to\infty} \nu^n \int_Q \frac{1}{\nu^n} F(p_0 + D\xi(y))dy \\ &= \int_Q F(p_0 + D\xi(y))dy \ , \end{aligned}$$

ξ is periodic.

Quasi-convexity is characterized (in the presence of additional strong assumptions) by the following two theorems which we state without proofs:

Theorem 1.2 (Morrey, Meyers [51], [53]) *If*

(i) $F \geq 0$

(ii) $|F(p) - F(q)| \leq k\left(1 + |p|^{s-1} + |q|^{s-1}\right)|p-q|, s \geq 1, 0 < k = const.$

then: $\mathbb{F}$ *is weakly lower semi-continuous* $\Leftrightarrow$ F *is quasi-convex.*

Theorem 1.3 (Acubi-Fusco [22]) *If*

(i) F *is a Caratheodory function*

(ii) $0 \leq F(x,u,p) \leq \lambda(1 + |u|^m + |p|^m)$, $m \geq 1$, $0 < \lambda = const$

then: F *is weakly lower semicontinuous* $\Leftrightarrow$ F *is quasiconvex*

Remarks:

1) Convexity implies quasi-convexity because $F(p_0 + D\xi) \geq F(p_0) + A(p)D\xi$.

2) F is called *weakly quasi-convex* iff $F(x_0, u_0, p^i_\alpha + \xi_\alpha \eta^i)$ is convex in ξ (for η fixed) respectively convex in η (for fixed ξ) i.e. convexity in certain directions. If $F \in C^2$ then:
F is weakly quasi-convex iff F satisfies a $L-H$-condition: $F_{p^i_\alpha p^j_\beta}\xi_\alpha\xi_\beta\eta^i\eta^j \geq 0$.
Quasi-convex implies weakly quasi-convex:
First observe that: F is quasi-convex iff F_ε (the mollified F) is quasi-convex

$$\begin{aligned}\int_\Omega F_\varepsilon(p + D\xi)dx &= \int_\Omega\int F(p - q + D\xi)\phi_\varepsilon(q)dqdx \\ &\geq \int F(p-q)\phi_\varepsilon(q)dq \\ &= \mathbb{F}_\varepsilon(p)\ .\end{aligned}$$

Linear functions are minimizers for $\mathbb{F}_\varepsilon$; therefore (by the same argument as on p. 13) we have a $L-H$-condition which implies weak quasi-convexity.

3) It is an open question if weakly quasi-convex implies quasi-convex. The conjecture is that it doesn't.

4) Examples of quasi-convex functions are the poly-convex functions of Ball. Recall that a *poly-convex function* is defined as a convex function of det Du, tr Du, etc.

Now if f is convex, $F(p) = f(\det Du)$ is quasi-convex:
$\det(p + D\phi) = \det p + \det D\phi + \langle p, D\phi\rangle$, where

$$\langle p, D\phi\rangle = p_1^1 D_2\phi^2 + p_2^2 D_1\phi^1 - p_2^1 D_1\phi^2 - p_1^1 D_2\phi^1 \qquad (\text{if } n = 2)$$

$$\begin{aligned}
&\int_\Omega f(\det(p + D\phi))dx \\
&\quad = \int_\Omega f(\det p + \det D\phi + \langle p, D\phi\rangle)dx \\
&\quad \geq \int_\Omega \left[f(\det p) + f_{p_\alpha^i}(\det p)(\det D\phi + \langle p, D\phi\rangle)\right] dx \\
&\quad = \int_\Omega f(\det p)dx + f_{p_\alpha^i}(\det p) \underbrace{\int_\Omega (\det D\phi + \langle p, D\phi\rangle)dx}_{\text{this term vanishes because } \phi \in C_0^\infty(\Omega)}
\end{aligned}$$

thus

$$\int_\Omega f(\det(p + D\phi))dx \geq f(\det p) \ .$$

For more information on the results of this section we refer to [1], [3], [4], [5], [35], [55], [60] and the references there.

Chapter 2

Differencequotient-method and linearisation of nonlinear systems; Hilbert-space regularity

2.1 Introduction

One of the fundamental tools in studying elliptic systems are *"a priori estimates"* and it is typical for such systems, that "a priori estimates" are also *"a posteriori estimates"*.

Another characteristic of estimates for elliptic systems is that they are *local.* Therefore in general one looks for local estimates in the interior (say on $B_R \subset\subset \Omega$) and local estimates on the boundary (= *boundary-estimates* on $B_R \cap \Omega (\neq \phi)$). From these one gains *global estimates* by a simple covering argument.

Interior local estimates for a solution u do not depend on the boundary values, but only on the values of u on a larger ball (say B_{2R}); whereas estimates near the boundary depend also on the boundary conditions.

Certainly one of the most important estimates is the energy-estimate, which can be regarded as the local version of coerciveness or of Gårding's inequality. It was explicitely stated in the thirties by Caccioppoli and Leray in the case of linear second order equations, but, as we shall see, it works in a suitable form for "all" linear and non-linear elliptic systems. It is the starting point for the regularity-theory and we shall refer to it as *Caccioppoli-inequality.*

Let us first consider the simplest case namely that of *harmonic functions:* Suppose $u \in H^1_{\text{loc}}(\Omega)$ (i.e. $u \in H^1(K)$ for every relatively compact subset K of Ω)

satisfies $\int_\Omega DuD\phi dx = 0$ for $\forall\phi \in H^1(\Omega)$ with supp $\phi \subset\subset \Omega$ (i.e. $\Delta u = 0$ in the sense of distributions).
Set $\phi := (u-\lambda)\eta^2$ where λ is a constant and η is a function with the following properties:

(i) $\eta \in C_0^\infty(B_R)$ (or $\eta \in \mathrm{Lip}(B_R)$)

(ii) $0 \leq \eta \leq 1$; $\eta \equiv 1$ in $B_\rho \subset B_R \subset\subset \Omega$

(iii) $|D\eta| \leq \frac{2}{R-\rho}$,

η is called *cut-off-function*.

We now have

$$\int_{B_R} Du\left\{(Du)\eta^2 - (u-\lambda)2\eta D\eta\right\} dx = 0$$

or

$$\begin{aligned}\int_{B_r} |Du|^2\eta^2 dx &= \left|\int_{B_r} DuDu\eta^2 dx\right| \\ &= \left|-2\int_{B_R} Du(u-\lambda)\eta D\eta dx\right| \\ &\leq 2\left(\int_{B_r} |Du|^2\eta^2 dx\right)^{1/2}\left(\int_{B_r} |u-\lambda|^2|D\eta|^2 dx\right)^{1/2}\end{aligned}$$

by dividing and squaring we get:

$$\int_{B_R} |Du|^2\eta^2 dx \leq 2\int_{B_r} |u-\lambda|^2|D\eta|^2 dx$$

and therefore (by the properties of η):

$$\int_{B_\rho} |Du|^2 dx \leq \frac{c}{(R-\rho)^2}\int_{B_r\setminus B_\rho} |u-\lambda|^2 dx\ .$$

(Observe that $\eta \equiv 1$ on B_ρ and hence that $D\eta \equiv 0$ on B_ρ.)

Set $\rho = R/2$ and $\lambda = 0$:

$$\int_{B_{R/2}} |Du|^2 dx \leq \frac{c}{R^2}\int_{B_R} |u|^2 dx \tag{2.1}$$

Set $\rho = R/2$ and $\lambda = \bar{u}_R := \fint_{B_R} u dx$:

$$\int_{B_{R/2}} |Du|^2 dx \leq \frac{c}{R^2} \int_{B_R} |u - \bar{u}_R|^2 dx \tag{2.2}$$

Set $\rho = R/2$ and $\lambda = \bar{u}_{R \setminus R/2}$:

$$\int_{B_{R/2}} |Du|^2 dx \leq \frac{c}{R^2} \int_{B_R \setminus R/2} |u - \bar{u}_{R \setminus R/2}|^2 dx \ . \tag{2.3}$$

(2.1), (2.2) and (2.3) are referred to as *Caccioppoli-inequalities.*

A priori- and a posteriori-estimates

Assume, that we already know, that $u \in C^\infty(\Omega)$ and satisfies

$$\int_\Omega DuD\phi dx = 0 \qquad \text{for } \forall \phi \in C_0^\infty(\Omega) \ .$$

Choose $\phi := D_s \psi$ where $\psi \in C_0^\infty(\Omega)$, then

$$\int_\Omega D(D_s u) D\phi dx = 0 \qquad \text{for } \forall \phi \in C_0^\infty(\Omega)$$

i.e. $D_s u$ is weakly harmonic.

We conclude that

$$\int_{B_{R/2}} |D^2 u|^2 dx \leq c(R) \int_{B_r} |u|^2 dx$$

and, by induction, for $k \in \mathbb{N}$:

$$\int_{B_{R/2}} |D^k u|^2 dx \leq c(R, k) \int_{B_R} |u|^2 dx$$

i.e. we have the a priori estimate

$$||u||_{H^k(B_{R/2})} \leq c(R, k) ||u||_{L^2(B_R)} \ . \tag{2.4}$$

Remark: If, instead of $u \in C^\infty(\Omega)$, we only assume $u \in H^1(\Omega)$, second derivatives have no meaning. But we can consider the ε-mollified of u:

$$u_\varepsilon(x) := \int_\Omega u(y) \psi_\varepsilon(x - y) dy = \int_\Omega u(x - y) \phi_\varepsilon(y) dy \ ,$$

where $\phi_\varepsilon(x) = \varepsilon^{-n}\psi\left(\frac{x}{\varepsilon}\right)$ and $\psi \geq 0$ supp $\psi \subset B_1$; $\fint_\Omega \psi dx = 1$. Then $u_\varepsilon \in C^\infty$ and we claim, that u_ε is in fact harmonic in $\Omega_\varepsilon = \{x \in \Omega \mid \operatorname{dist}(x, \partial\Omega) > \varepsilon\}$:

$$\int_{\Omega_\varepsilon} D_x \int u(x-y)\phi_\varepsilon(y)dy D_x\phi dx = 0 \qquad \text{for } \forall\phi \in C_0^\infty(\Omega_\varepsilon)$$

and

$$\begin{aligned} 0 &= \int\int D_x u(x-y)\phi_\varepsilon(y)D_x\phi dxdy \\ &= \int\int D_z u(z)\phi_\varepsilon(x-z)D_z\phi/z+y)dzdy \\ &= \int_{\Omega_\varepsilon} DuD\phi_\varepsilon dx \end{aligned}$$

hence we can control all derivatives of u_ε by the L^2-norm of u and because $u_\varepsilon \xrightarrow{L^2} u$ the same holds for u; i.e. (2.4) is also a regularity result (an *"a posteriori-estimate"*).

We will not consider this construction further but follow another method known as the *difference-quotient-technique* of Nirenberg [63] (see 2.3 for details!):

We define $\tau_{h,s}u(x) := \frac{1}{h}(u(x+he_s) - u(x))$ with

$$e_s = (0, 0, \ldots, 0, \underset{\substack{\uparrow \\ s}}{1}, 0, \ldots, 0)$$

$\int_\Omega D(\tau_{h,s}u)D\phi dx = 0$ follows from $\int_\Omega DuD\phi dx = 0$ simply by taking the difference. So by (2.1):

$$\int_{B_{R/2}} |D(\tau_{h,s}u)|^2 dx \leq c(R) \int_{B_R} |\tau_{h,s}u|^2 dx \ .$$

One might expect that this implies

$$||D^2u||_{L^2(B_{R/2})} \leq c(R)||u||_{L^2(B_R)}$$

which in fact is true, as we shall see in 2.3. But let us first look at two other consequences of the Caccioppoli-inequalities:

1) We use the Poincaré-inequality

$$\int_{B_R \setminus B_{R/2}} \left|u - \bar{u}_{R\setminus R/2}\right|^2 dx \leq c(n)R^2 \int_{B_R \setminus B_{R/2}} |Du|^2 dx \ .$$

On both sides we add $c(n) \int\limits_{B_{R/2}} |Du|^2 dx$ (this is sometimes referred to as the "hole-filling"-technique of Widman) and we get:

$$\int\limits_{B_{R/2}} |Du|^2 dx \le \underbrace{\frac{c(n)}{1+c(n)}}_{<1} \int\limits_{B_r} |Du|^2 dx$$

now we take the limit $R \to \infty$:

$$\int\limits_{\mathbb{R}^n} |Du|^2 dx \le \theta \int\limits_{\mathbb{R}^n} |Du|^2 dx \qquad \text{with } \theta < 1$$

and we conclude:
if $\int\limits_{\mathbb{R}^n} |Du|^2 dx < \infty$ then $Du \equiv 0$, i.e. $u =$ const.

We have also: *A bounded harmonic function in $\mathbb{R}^2$ must be constant.*

In fact, since $|u| \le M$

$$\int\limits_{B_R} |Du|^2 dx \le \frac{c}{R^2} \int\limits_{B_{2R}} |u|^2 dx \le \tilde{c}$$

where $\tilde{c}$ is a constant independent of R, then by the above argument $Du \equiv 0$, i.e. $u =$ const.

2) We recall the *Sobolev-Poincaré-inequality*

$$\left(\int\limits_{B_r} |u - \bar{u}_R|^{p^*} dx \right)^{1/p^*} \overset{\text{Sobolev}}{\le} c(n,p) \Bigg(\int\limits_{B_R} |Du|^p dx +$$

$$+ R^{-p} \int\limits_{B_R} |u - \bar{u}_R|^p dx \Bigg)^{1/p} \overset{\text{Poincaré}}{\le} c(n,p) \left(\int\limits_{B_R} |Du|^p dx \right)^{1/p}$$

where $p^* = \frac{np}{n-p} > p$ is the Sobolev-exponent.

With this inequality and (2.2) we have (set $p^* = 2$ then $p = \frac{2n}{2+n} < p^*$)

$$\int\limits_{B_{R/2}} |Du|^2 dx \le \frac{c}{R^2} \left(\int\limits_{B_R} |Du|^p dx \right)^{2/p} .$$

Dividing by R^n we find

$$\left(\fint_{B_{R/2}} |Du|^2 dx \right)^{1/2} \leq c \left(\fint_{B_R} |Du|^p dx \right)^{1/p}$$

where $p < 2$ and c is indepenent of R.

This is a *Hölder-reverse-inequality with increasing support* which is important in the theory of nonlinear elliptic systems. We will come back to it later.

2.2 The Caccioppoli-inequality

Let us look at systems

$$(*) \qquad -D_\alpha \left(A_{ij}^{\alpha\beta}(x) D_\beta u^j \right) = f_i - D_\alpha f_i^\alpha, \text{ for } i = 1, 2, \ldots, N,$$

where we assume that $f_i, f_i^\alpha \in L^2(\Omega)$.

Suppose now that $u \in H^{1,2}_{\text{loc}}(\Omega, \mathbb{R}^N)$ is a solution of $(*)$, i.e.

$$\int_\Omega A_{ij}^{\alpha\beta}(x) D_\beta u^j D_\alpha \phi^i dx = \int_\Omega (f_i \phi^i + f_i^\alpha D_\alpha \phi^i) dx \qquad \text{for } \forall \phi \in C_0^\infty(\Omega) \ .$$

If we assume

$$\begin{cases} 1) & A_{ij}^{\alpha\beta} \in L^\infty(\Omega) \\ 2) & A_{ij}^{\alpha\beta} \xi_\alpha^i \xi_\beta^j \geq \nu |\xi|^2, \nu > 0 \text{ (strong ellipticity)} \end{cases}$$

or

$$\begin{cases} 1) & A_{ij}^{\alpha\beta} = \text{ const.} \\ 2) & \text{Legendre-Hadamard-condition holds:} \\ & A_{ij}^{\alpha\beta} \xi^i \xi^j \eta_\alpha \eta_\beta \geq \nu |\xi|^2 |\eta|^2, \nu > 0 \text{ for all } \xi \in \mathbb{R}^N \text{ and all } \eta \in \mathbb{R}^n. \end{cases}$$

or

$$\begin{cases} 1) & A_{ij}^{\alpha\beta} \in C^0(\Omega) \\ 2) & \text{Legendre-Hadamard-condition holds} \\ 3) & R \text{ small } (R < R_0 \text{ say}). \end{cases}$$

then we have the Cacioppoli-inequality:

$$\int_{B_{R/2}} |Du|^2 dx \leq c(n) \left\{ \frac{1}{R^2} \int_{B_R} |u - \lambda|^2 + R^2 \int_{B_R} \sum_i f_i^2 dx + \int_{B_R} \sum_{i,\alpha} (f_i^\alpha)^2 dx \right\}$$

We give a proof for the third case:
From

$$\int_{B_R(x_0)} A_{ij}^{\alpha\beta}(x_0) D_\alpha u^i D_\beta \phi^j dx = \int_{B_R(x_0)} \left[A_{ij}^{\alpha\beta}(x_0) - A_{ij}^{\alpha\beta}(x)\right] D_\alpha u^i D_\beta \phi^j dx + \int_{B_R(x_0)} f_i \phi^i dx + \int_{B_R(x_0)} f_i^\alpha D_\alpha \phi^i dx \quad \text{for}$$

$\forall \in H_0^1(B_R)$ choosing $\phi = (u - \lambda)\eta^2$ as before in 2.1 we deduce (compare with the proof of Gårding's inequality):

$$\begin{aligned}
&\int_{B_R(x_0)} |D[(u-\lambda)\eta]|^2 dx \\
\leq &\int_{B_R(x_0)} A_{ij}^{\alpha\beta}(x_0) D_\alpha \left[(u^i - \lambda^i)\eta\right] D_\beta \left[(u^j - \lambda^j)\eta\right] dx \\
= &\int_{B_R(x_0)} A_{ij}^{\alpha\beta}(x_0) D_\alpha u^i D_\beta \left[(u^j - \lambda^j)\eta^2\right] \\
&+ \int_{B_R(x_0)} A_{ij}^{\alpha\beta}(x_0)(u^i - \lambda^i)(u^j - \lambda^j) D_\alpha \eta D_\beta \eta dx \\
= &\int_{B_R(x_0)} \left[A_{ij}^{\alpha\beta}(x_0) - A_{ij}^{\alpha\beta}\right] D_\alpha u^i D_\beta \left[(u^j - \lambda^j)\eta^2\right] dx + \int_{B_R(x_0)} f_i (u^i - \lambda^i)\eta^2 dx \\
&+ \int_{B_R(x_0)} f_i^\alpha D_\alpha \left[(u^i - \lambda^i)\eta^2\right] dx + \int_{B_R(x_0)} A_{ij}^{\alpha\beta}(x_0)(u^i - \lambda^i)(u^j - \lambda^j) D_\alpha \eta D_\beta \eta dx \\
\leq &\ \delta \int_{B_R(x_0)} D_\alpha u^i D_\beta u^j \eta^2 + 2\delta \int_{B_R(x_0)} D_\alpha u^i D_\beta \eta (u^j - \lambda^j)\eta dx \\
&+ \int_{B_R(x_0)} f_i \eta (u^i - \lambda^i) dx + \int_{B_R(x_0)} f_i^\alpha D_\alpha u^i \eta^2 dx \\
&+2 \int_{B_R(x_0)} \sum f_i^\alpha (u^i - \lambda^i)\eta D\eta dx \\
\leq &\ \delta \int_{B_R(x_0)} |Du|^2 \eta^2 dx + 2\delta\varepsilon \int_{B_R(x_0)} |Du|^2 \eta^2 dx + 2\varepsilon^{-1}\delta \int_{B_R(x_0)} |u-\lambda|^2 |D\eta|^2 dx \\
&+R^{-2} \int_{B_R(x_0)} |u-\lambda|^2 dx + R^2 \int_{B_R(x_0)} \sum f_i^2 dx + \varepsilon^{-1} \int_{B_R(x_0)} \sum (f_i^\alpha) dx
\end{aligned}$$

$$+\varepsilon \int_{B_R(x_0)} |Du|^2\eta^2 + 2\varepsilon^{-1} \int_{B_R(x_0)} \sum (f_i^\alpha)^2 dx + 2\varepsilon \int_{B_R(x_0)} |u-\lambda|^2 |D\eta|^2 dx$$

$$\leq (\delta + 2\delta\varepsilon + \varepsilon) \int_{B_R(x_0)} |Du|^2\eta^2 dx + 3\varepsilon^{-1} \int_{B_R(x_0)} (f_i^\alpha)^2 dx$$

$$+R^{-2}(1 + 2\varepsilon^{-1}\delta + 2\varepsilon) \int_{B_R(x_0)} |u-\lambda|^2 dx + R^2 \int_{B_R(x_0)} \sum f_i^2 dx$$

furthermore one has

$$\begin{aligned}
\int_{B_{R/2}(x_0)} |Du|^2 dx &\leq \int_{B_R(x_0)} |Du|^2 R^2 dx \\
&\leq \int_{B_R(x_0)} |D[(u-\lambda)\eta]|^2 + 2\varepsilon \int_{B_R(x_0)} |Du|^2\eta^2 dx \\
&\quad +(2\varepsilon^{-1} + 1) \int_{B_R(x_0)} |u-\lambda|^2 |D\eta|^2 dx
\end{aligned}$$

so that

$$(1-2\varepsilon) \int_{B_{R/2}(x_0)} |Du|^2 dx \leq \int_{B_R(x_0)} |D[(u-\lambda)\eta]|^2 dx + R^{-2}(1+2\varepsilon^{-1}) \int_{B_R(x_0)} |u-\lambda|^2 dx \ .$$

Inserting the above estimate in the last one we get the result because ε is arbitrary and δ can be chosen sufficiently small.

2.3 The difference-quotient-method

We recall the definition of $\tau_{h,s}$: for u defined in Ω we set

$$(\tau_{h,s}u)(x) := \frac{1}{h}\left(u(x + \text{he}_s) - u(x)\right) \quad \text{where} \quad (0,\dots,0,\underset{\substack{\uparrow \\ s}}{1},0,\dots,0)$$

$\tau_{h,s}$ is defined in $\Omega_h = \{x \in \Omega \mid \text{dist}(x, \partial\Omega) < h\}$ or more precisely in $\Omega_{s,h} = \{x \in \Omega | x + \text{he}_s \in \Omega\}$. We list some elementary properties of $\tau_{h,s}$:

1) If $u \in H^{1,p}(\Omega)$ then $\tau_{h,s}u \in H^{1,p}(\Omega_{s,h}), \tau_{h,s}Du = D\tau_{h,s}u$.

2) If either u or v have compact support then

$$\int_\Omega u\tau_{h,s}v dx = -\int_\Omega v\tau_{-h,s}u dx \ .$$

3) $\tau_{h,s}(uv)(x) = u(x + \mathrm{he}_s)\tau_{h,s}v + v(x)\tau_{h,s}u$.

Proposition 2.1 *(i) If $\tilde{\Omega} \subset\subset \Omega$ (i.e. $\tilde{\Omega}$ strictly contained in Ω) there exists a constant $K(\Omega, \tilde{\Omega})$ such that*

$$||\tau_{h,s}v||_{L^p(\tilde{\Omega})} \leq K(\Omega, \tilde{\Omega})||D_s v||_{L^p(\Omega)}$$

for $\forall v \in H^{1,p}(\Omega)$ and h sufficiently small.

(ii) If $v \in L^p(\Omega)$ and $||\tau_{h,s}v||_{L^p(\tilde{\Omega})} \leq k$ (= constant independent of h) then $v \in H^{1,p}(\tilde{\Omega})$, $||D_s v||_{L^p(\tilde{\Omega})} \leq k$ and $\tau_{h,s}v \xrightarrow{L^p(\tilde{\Omega})} D_s v$ (for $h \to 0$).

PROOF:

(i) We may assume that $\tilde{\Omega}$ = cube Q in $\mathbb{R}^n$; $s = n$, $\tau_h := \tau_{h,n}$ (By covering $\tilde{\Omega}$ by cubes the general result follows!)

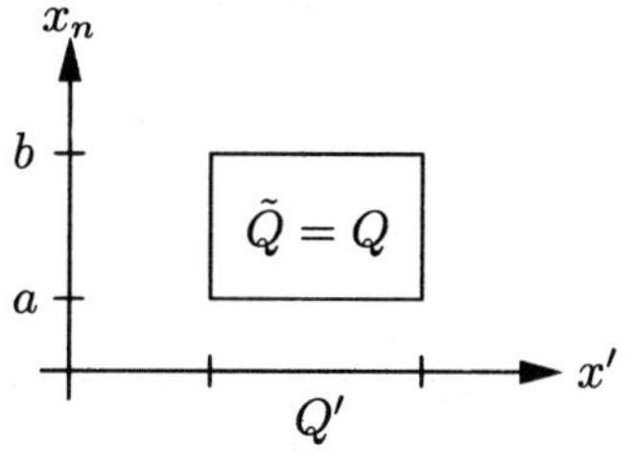

We use the fact that functions in Sobolev-spaces are absolutely continuous for a.e. line; hence

$$\tau_h(v)(x) = \frac{1}{h}\int_{x_n}^{x_n+h} D_n v(x', t)dt \qquad \text{for a.e. } x$$

and

$$\begin{aligned} \int_{\tilde{\Omega}} |\tau_h v|^p dx &= \int_a^b dx_n \int_{Q'} dx' \left| \frac{1}{h} \int_{x_n}^{x_n+h} D_n v(x', t)dt \right|^p \\ &\leq \int_{Q'} dx' \int_{a-h_0}^{b+h_0} |D_n v(x', x_n)|^p \, dx_n \\ &\leq \int_{\Omega} |Dv|^p dx \end{aligned}$$

where we used Hardy's inequality:

$$\begin{aligned}
\int_a^b \left| \frac{1}{h} \int_x^{x+h} f(t)dt \right|^p dx &= \int_a^b \left| \frac{1}{h} \int_0^h f(x+t)dt \right|^p dx \\
&\leq \int_a^b \frac{1}{h} \int_0^h |f(x+t)|^p dt dx \\
&= \frac{1}{h} \int_0^h dt \int_a^b |f(x+t)|^p dx \\
&\leq \int_{a-h_0}^{b+h_0} |f|^p dx \ .
\end{aligned}$$

(ii) We can select a subsequence

$$\tau_{h_j,s} v \longrightarrow g \qquad \text{in } L^p(\tilde{\Omega})$$

then $g = D_s v$ in the sense of distributions:

$$\int_{\tilde{\Omega}} g\phi dx = \lim_{j\to\infty} \int_{\tilde{\Omega}} \tau_{h_j,s} v \phi dx = - \lim_{j\to\infty} \int_{\tilde{\Omega}} v \tau_{-h_j,s} \phi dx = - \int_{\tilde{\Omega}} v D_s \phi dx$$

for $\forall \phi \in C_0^\infty(\tilde{\Omega})$,

By the uniqueness of the distributional derivative we have moreover, that the whole sequence $\tau_{h,s} v$ weakly converges to $D_s v$ and by semicontinuity

$$\int_{\tilde{\Omega}} |D_s v|^p dx \leq \liminf_{h\to 0} \int_{\tilde{\Omega}} |\tau_{h,s} v|^p dx \ .$$

It remains to show the strong convergence:

$$\begin{aligned}
\int_{\tilde{\Omega}} |\tau_{h,s} v(x) - D_s v(x)|^p dx &= \int_{\tilde{\Omega}} \left| \frac{1}{h} \int_0^h [D_s v(x + \mathrm{t} e_s) - D_s v(x)]\, dt \right|^p dx \\
&\leq \int_{\tilde{\Omega}} \frac{1}{h} \int_0^h |D_s v(x + \mathrm{t} e_s) - D_s v(x)|^p \, dt dx \\
&= \frac{1}{h} \int_0^h dt \int_{\tilde{\Omega}} |D_s v(x + \mathrm{t} e_s) - D_s v(x)|^p \, dx
\end{aligned}$$

and this converges to zero as h goes to zero.

2.4 Hilbert-space regularity: interior regularity

Let us consider the linear elliptic system

$$\int_\Omega A_{ij}^{\alpha\beta}(x) D_\alpha u^i D_\beta \phi^j dx = \int_\Omega (f_i \phi^i + f_i^\alpha D_\alpha \phi)^i dx \forall \phi \in H_0^1(\Omega, \mathbb{R}^N) \tag{2.5}$$

where $A_{ij}^{\alpha\beta} \in \mathrm{Lip}(\Omega)$, $f = (f_i) \in L^2(\Omega, \mathbb{R}^N)$ and $F = (f_i^\alpha) \in H^{1,2}(\Omega, \mathbb{R}^{nN})$. For the sake of simplicity we shall suppose that (2.5) is elliptic in the sense that

$$A_{ij}^{\alpha\beta} \xi_\alpha^i \xi_\beta^j \geq |\xi|^2 \qquad \text{for } \forall \xi \in \mathbb{R}^{nN} .$$

But, as will become clear from the proofs, it would be sufficient to assume that the Legendre-Hadamard-condition holds, i.e.

$$A_{ij}^{\alpha\beta} \xi^i \xi^j \eta_\alpha \eta_\beta \geq |\xi|^2 |\eta|^2 \qquad \text{for } \forall \xi \in \mathbb{R}^N \text{ and } \forall \eta \in \mathbb{R}^n .$$

We have

$$\int_\Omega A(x+\mathrm{he}_s) Du(x+\mathrm{he}_s) D\phi(x) dx = \int_\Omega f(x+\mathrm{he}_s)\phi(x) dx + \int_\Omega F(x+\mathrm{he}_s) D\phi(x) dx$$

subtract (2.5) and divide by h to get

$$\int_\Omega A(x+\mathrm{he}_s) \left[\frac{1}{h} (Du(x+\mathrm{he}_s) - Du(x)) \right] D\phi dx \tag{2.6}$$

$$+ \int_\Omega \frac{1}{h} \left[A(x+\mathrm{he}_s) - A(x) \right] Du D\phi dx$$

$$= \int_\Omega \frac{1}{h} (f(x+\mathrm{he}_s) - f(x)) \phi(x) dx + \int_\Omega \frac{1}{h} (F(x+\mathrm{he}_s) - F(x)) D\phi(x) dx . \tag{2.7}$$

Suppose first $f \equiv 0$ then by Caccioppoli's inequality

$$\int_{B_{R/2}} |D\tau_h u|^2 dx \leq c(R) \left(\frac{1}{R^2} \int_{B_R} |\tau_h u|^2 dx + \int_{B_R} |\tau_h A|^2 |Du|^2 dx + \int_{B_R} |\tau_h F|^2 dx \right) \tag{2.8}$$

the right hand side is uniformly bounded and therefore we conclude by (ii) of proposition 2.1 that the second derivative exists in $L^2_{\mathrm{loc}}(\Omega)$.

If $f \neq 0$ the term (2.6) above can be treated as in the proof of Caccioppoli's inequality:

We have

$$\int_\Omega \tau_h f \phi dx = -\int_\Omega f \tau_{-h} \phi dx$$

set $\phi = \eta^2 \tau_h u$ (η as before the cut-off-function), then

$$\int_\Omega f\tau_{-h}(\eta^2\tau_h u)dx = \int_\Omega f\eta\tau_h u \tau_{-h}\eta dx + \int_\Omega f\eta(x - \mathrm{he}_s)\tau_{-h}(\eta\tau_h u)dx$$

where the second term is bounded by

$$\frac{1}{\varepsilon}\int_\Omega |f\eta|^2 dx + \varepsilon\left[\int_\Omega \eta^2|D\tau_h u|^2 dx + \int_\Omega |D_\eta|^2|\tau_h u|^2 dx\right]$$

(this term can be brought to the left hand side in (2.8)).

We have proved the

Theorem 2.1 *Suppose that* $A_{ij}^{\alpha\beta} \in \mathrm{Lip}(\Omega)$, *that (2.5) is elliptic and that* $f_i \in L^2(\Omega)$ *and* $f_i^\alpha \in H^{1,2}(\Omega)$. *If* $u \in H^{1,2}_{\mathrm{loc}}(\Omega, \mathbb{R}^N)$ *is a weak solution of the elliptic system*

$$-D_\alpha\left(A_{ij}^{\alpha\beta}(x)D_\beta u^j\right) = f_i - D_\alpha f_i^\alpha \qquad \text{for } j = 1, \dots, N$$

then $u \in H^{2,2}_{\mathrm{loc}}(\Omega, \mathbb{R}^N)$.

The procedure can be continued in the following way:
Set $\phi = D_s\psi$ for $\psi \in C_0^\infty(\Omega)$ in

$$\int_\Omega ADuD\phi dx = \int_\Omega f\phi dx + \int_\Omega FD\phi dx$$

then

$$-\int_\Omega D_s(ADu)D\psi dx = \int_\Omega fD_s\psi dx - \int_\Omega D_sFD\phi dx$$

or

$$\begin{aligned}\int_\Omega AD(D_su)D\psi dx &= -\int_\Omega D_sADuD\psi dx - \int_\Omega fD_s\psi dx + \int_\Omega D_sFD\phi dx \\ &= \int_\Omega [-D_sADu - f - D_sF]D\psi dx\end{aligned}$$

the same considerations as above together with the assumptions $D_sA \in \mathrm{Lip}(\Omega)$, $f_i \in H^1(\Omega)$, $f_i^\alpha \in H^2(\Omega)$ lead to the result $D_su \in H^{2,2}_{\mathrm{loc}}(\Omega)$ or $u \in H^{3,2}_{\mathrm{loc}}(\Omega)$.

By induction we then get

Theorem 2.2 *Suppose u is a weak solution of the elliptic system*

$$-D_\alpha(A_{ij}^{\alpha\beta}D_\beta u^j) = f_i - D_\alpha f_i^\alpha \qquad \text{for} j = 1,\dots,N$$

with

$$\begin{array}{rcl} A_{ij}^{\alpha\beta} & \in & C^{k,1}(\Omega) \\ f_i & \in & H^{k,2}(\Omega) \\ f_i^\alpha & \in & H^{k+1,2}(\Omega) \end{array}$$

then

$$u \in H^{k+2,2}_{\text{loc}}(\Omega, \mathbb{R}^N) .$$

Remark: Note that if $A_{ij}^{\alpha\beta}$, f_i, $f_i^\alpha \in C^\infty(\Omega)$, then $u \in C^\infty(\Omega)$; this is in particular true if $A_{ij}^{\alpha\beta}=$ const., $f_i \equiv 0$ and $f_i^\alpha \equiv 0$. In this case (more precisely) we have for any ball $B_R \subset\subset \Omega$ and for any k

$$||u||_{H^{k,2}(B_{R/2},\mathbb{R}^N)} \leq c(k,R)||u||_{L^2(B_R,\mathbb{R}^N)} .$$

2.5 Hilbert-space regularity: boundary regularity

In this section we consider the Dirichlet-problem:

$$\int_\Omega ADuD\phi dx = \int_\Omega f\phi dx + \int_\Omega FD\phi dx, \qquad u = u_0 \text{ on } \partial\Omega . \tag{2.9}$$

We set $w := u - u_0$ then we satisfy a system of the same kind, but we have reduced the boundary condition to the homogeneous one. So we may assume that $u_0 = 0$ in (2.9). Furthermore we assume that $\partial\Omega \in C^{k+1}$. This means that for any $x_0 \in \partial\Omega$ there exists a diffeomorphism

$$G : B_R^+ \to M \cap \Omega \quad \text{of class } C^{k+1}$$

which maps Γ_R onto $M \cap \partial\Omega$ (here B_R^+ is a halfball and M a neighbourhood of x_0)

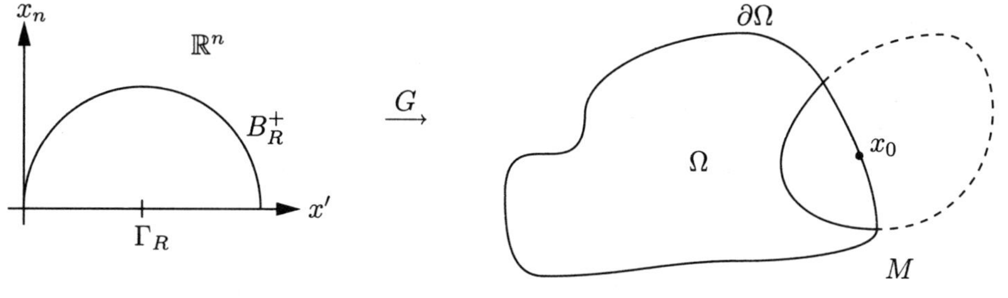

The diffeomorphism G induces a map G_* defined by $(G_*u)(y) := u(G(y))$, $y \in B_R^+$. G_* is an isomorphism between $H^{k+1}(B_R^+)$ and $H^{k+1}(\Omega \cap M)$.

We set $U := G_*u$ then $Du(G(y)) = J_G^{-1}DU(y)$ where J_G is the Jacobian-matrix of G. U satisfies the equation

$$\int\limits_{B_R^+} \tilde{A}DUD\phi dy = \int\limits_{B_R^+} \tilde{f}\phi dy + \int\limits_{B_R^+} \tilde{F}D\phi dy \tag{2.10}$$

with

$$\begin{aligned} \tilde{A} &= |\det J_G| J_G^{-1} AJ \\ \tilde{F} &= |\det J_G| J_G^{-1} G_* f_i^\alpha \\ \tilde{f} &= |\det J_G| G_* f \end{aligned}$$

clearly $\tilde{A}$ is also elliptic.

We have thus locally reduced the problem to $U = 0$ on Γ_R. If we now, as in the interior, take $\Phi = \tau_{-h,s}(\eta^2 \tau_{h,s} U)$ for $s = 1, 2, 3, \ldots, n-1$ and $\eta \in D_0^\infty(B_R(x_0))$ we deduce (exactly as in the interior case)

$$||\tau_{h,s} DU||_{L^2(B^+_{R/2})} \leq c\left\{||U_x||_{L^2(B_R^+)} + ||f||_{L^2(B_R^+)} + ||F_x||_{L^2(B_R^+)}\right\}$$

i.e., we can estimate all the second derivatives except $U^i_{x_n x_n}$.

From equation (2.10) we have

$$\int\limits_{B_R^+} \tilde{A}_{ij}^{nn}(x) D_n U^i D_n \phi^j dy = -\sum_{\alpha,\beta=1}^{n-1} \int\limits_{B_R^+} \tilde{A}_{ij}^{\alpha\beta} D_\alpha U^i D_\beta \phi^j dy + \int\limits_{B_R^+} \tilde{f}_i \phi^i dy + \int\limits_{B_R^+} F_i^\alpha D_\alpha \phi^i dy$$

so we see that $U^i_{x_n x_n}$ exist also in $L^2(B_{R^+})$. By induction and a covering argument we finally get

Theorem 2.3 *Suppose that $\partial\Omega \in C^{k+1}$, $A_{ij}^{\alpha\beta} \in C^k(\Omega)$, $f_i \in H^{k-1}(\Omega)$ and $f_i^\alpha \in H^k(\Omega)$. Let u be a solution of the (homogeneous) Dirichlet-problem then $u \in H^{k+1}(\Omega)$ and moreover*

$$||u||_{H^{k+1}(\Omega,\mathbb{R}^N)} \leq c(\Omega)\left\{||f||_{H^{k-1}(\Omega,\mathbb{R}^N)} + ||F||_{H^k(\Omega,\mathbb{R}^{nN})} + ||u||_{L^2(\Omega,\mathbb{R}^N)}\right\} .$$

In particular ($k = 1$) if $A_{ij}^{\alpha\beta}$= const. and u is a solution of

$$D_\alpha(A_{ij}^{\alpha\beta}) = f^i \quad \text{in} \quad \Omega\ , \qquad \text{for } j = 1, \ldots, N, u = 0 \text{ on } \partial\Omega$$

then

$$||u_{xx}||_{L^2(\Omega,\mathbb{R}^N)} \leq c(\Omega)||f||_{L^2(\Omega,\mathbb{R}^N)} .$$

In fact

$$||u||_{L^2(\Omega,\mathbb{R}^n)} \leq c||f||_{L^2(\Omega,\mathbb{R}^N)} .$$

2.6 Linearization of nonlinear equations

For $F \in C^2(\Omega)$ (or $\in C^{1,1}(\Omega)$) with

$$|F_p(p)| \leq L|p| \,, \quad |p|^2 \leq F(p) \leq \mu |p|^2$$

$$F_{p^i_\alpha p^j_\beta}(p)\xi^i_\alpha \xi^j_\beta \geq \nu|\xi|^2, \quad F_{pp}(p) \leq L \qquad (L, \mu, \nu \text{ are positive constants})$$

let us look at the functional $\int_\Omega F(Du)dx$.

If u is a minimizer in $H^{1,2}_{\text{loc}}(\Omega)$ then

$$\int_\Omega F_{p^i_\alpha}(Du)D_\alpha \phi^i dx = 0 \qquad \text{for } \forall \phi \in H^{1,2}_0(\Omega) \ .$$

More generally we consider weak solutions $u \in H^{1,2}_{\text{loc}}(\Omega)$ of

$$\int_\Omega A^\alpha_i(Du)D_\alpha \phi^i dx = 0 \qquad \text{for } \forall \phi \in H^{1,2}_0(\Omega)$$

where the A^α_i satisfy the growth conditions

$$\begin{aligned} \left|A^\alpha_{ip^j_\beta}(p)\right| &\leq M = \text{const.} \\ A^\alpha_{ip^j_\beta}\xi^i_\alpha \xi^j_\beta &\geq \nu|\xi|^2, \ \nu > 0 \,, \qquad \text{for } \forall \xi \in \mathbb{R}^N \ . \end{aligned}$$

Then we have

$$\int_\Omega \left[A^\alpha_i\left(Du(x + \text{he}_s)\right) - A^\alpha_i(Du(x))\right] D_\alpha \phi^i dx = 0$$

and for a.e. $x \in \Omega$

$$\begin{aligned} &A^\alpha_i(Du(x + \text{he}_s)) - A^\alpha_i(Du(x) \\ &= \int_0^1 \frac{d}{dt} A^\alpha_i\left((tDu(x + \text{he}_s) + (1-t)Du(x)\right) dt \\ &= \int_0^1 A^\alpha_{ip^j_\beta}\left(tDu(x + \text{he}_s) + (1-t)DU(x)dt \cdot D_\beta \left(u^i(x + \text{he}_s) - u^i(x)\right)\right. \end{aligned}$$

hence, writing

$$\tilde{A}^{\alpha\beta}_{ij}(x,h) := \int_0^1 A^\alpha_{ip^j_\beta}\left(tDu(x + \text{he}_s) + (1-t)Du(x)\right) dt$$

we get

$$\int_\Omega \tilde{A}_{ij}^{\alpha\beta}(x,h)D_\beta(\tau_h u^j)D_\alpha\phi^i dx = 0 \qquad \text{for } \forall\phi \in C_0^\infty(\Omega)$$

which implies the existence of the second derivatives of u in $L^2(\Omega)$ as in the linear case (Caccioppoli inequality!).

Passing to the limit $h \to 0$ we find

$$\int_\Omega A^\alpha_{ip^j_\beta}(Du)D_\beta(Du)D_\alpha\phi dx = 0 \qquad \text{for } \forall\phi \in H_0^1(\Omega, \mathbb{R}^{nN}) \ . \tag{2.11}$$

(2.11) can be rewritten as a *quasilinear system* for the vector-valued function $U := (D_s u^j)$:

$$\int_\Omega \delta_{\ell s} A^\alpha_{ip^j_\beta}(U)D_\beta U^j_s D_\alpha\phi^i_\ell dx = 0 \qquad \text{for } \forall\phi \in H_0^1(\Omega, \mathbb{R}^{nN}) \ . \tag{2.12}$$

We have proved the

Theorem 2.4 *Let $u \in H^1(\Omega, \mathbb{R}^N)$ be a solution of*

$$\int_\Omega A_i^\alpha(Du)D_\alpha\phi^i dx = 0 \qquad \text{for } \forall\phi \in H_0^1(\Omega, \mathbb{R}^N)$$

then $u \in H^{2,2}_{\text{loc}}(\Omega, \mathbb{R}^N)$ and the derivatives of u satisfy the quasilinear system (2.12).

System (2.12) is of the form

$$\int_\Omega A_i^\alpha(u)D_\beta u D_\alpha\phi dx = 0 \ . \tag{2.13}$$

Now it is clear that we cannot continue in the procedure of differentiating, since then we would get

$$0 = \int_\Omega A(u)D_\beta(D_s u)D_\alpha\phi dx + \int_\Omega A_{u^\ell}(u)\underbrace{D_s u^\ell D_\beta u}_{\in L^1(\Omega)} D_\alpha\phi dx$$

and the second term is uncontrollable.

Actually, the result is not even true; in general solutions of (2.13) are *not* in $H^{2,2}_{\text{loc}}(\Omega)$. But we have the following

Theorem 2.5 *Let $u \in H^{1,2}_{\text{loc}}(\Omega)$ be a solution of*

$$\int_\Omega A_i^\alpha(u, Du) D_\alpha \phi^i dx = 0 \qquad \text{for } \forall \phi \in C_0^\infty(\Omega)$$

with

$$\begin{aligned} |A_i^\alpha(u, Du)| &\leq c|Du| \\ \left|A^\alpha_{ip^j_\beta}(u, Du)\right| &\leq M \\ A^\alpha_{ip^j_\beta} \xi^i_\alpha \xi^j_\beta &\geq \nu|\xi|^2 \\ \left|A^\alpha_{iu^\ell}(u, Du)\right| &\leq c|Du| \end{aligned}$$

for some constants c, M, ν. If u is also continuous then $u \in H^{2,2}_{\text{loc}}(\Omega)$.

Hint for the proof: Use as test functions

$$\Phi = D_s(D_s u \eta^2)$$

and

$$\Phi = (u - \bar{u}_R)|Du|^2 \eta^2 \, !$$

For more information on the topics of this chapter we refer to [1], [5], [22], [55], [60].

Chapter 3

Schauder-estimates

In this chapter we prove Schauder-type estimates for solutions of linear elliptic systems with continuous (or Hölder-continuous) coefficients. These are estimates of the Hölder-norm of the derivatives of solutions.

We consider elliptic systems both in divergence and nondivergence form and prove Hölder-estimates in the interior of some domain Ω as well as estimates up to the boundary $\partial\Omega$ (Dirichlet-boundary condition).

Schauder-estimates are actually well known; but we shall present here a simple derivation without use of potential theory. It is based on ideas of C.B. Morrey [54] and S. Campanato [8] and was carried out essentially in [8].

For more information concerning the literature we refer the interested reader to [22].

3.1 Morrey- and Campanato-spaces

Let Ω be a *bounded* domain in $\mathbb{R}^n$. We shall assume that for all $x \in \Omega$ and for all $\rho \leq \operatorname{diam}\Omega$

(*) $\qquad \operatorname{meas}(B_\rho(x) \cap \Omega) \geq A\rho^n$. ($A$ is a positive constant).

This means that Ω cannot have "sharp outward cusps". For example Lipschitz-domains satisfy condition (*) .

We first define the *Morrey-spaces* $L^{p,\lambda}$: For real number $p \geq 1$ and $\lambda \geq 0$ set

$$L^{p,\lambda}(\Omega) := \left\{ u \in L^p(\Omega) \middle| \sup_{\substack{x_0 \in \Omega \\ 0<\rho<\operatorname{diam}\Omega}} \rho^{-\lambda} \int_{\Omega(x_0,\rho)} |u|^p dx < \infty \right\}$$

where $\Omega(x_0, \rho) = B(x_0, \rho) \cap \Omega$ and $B(x_0, \rho)$ is the ball of radius ρ centered at x_0.

Remarks:

1) Set $||u||^p_{L^{p,\lambda}(\Omega)} := \sup_{\substack{x_0\in\Omega \\ 0<\rho<\text{diam}\Omega}} \rho^{-\lambda} \int_{\Omega(x_0,\rho)} |u|^p dx$ then with respect to this norm $L^{p,\lambda}(\Omega)$ is a Banach-space.

2) $u \in L^{p,\lambda}$ iff $\int_{\Omega(x_0,\rho)} |u|^p dx \le c\rho^\lambda$ for all $x_0 \in \Omega$ and all ρ with $0 < \rho < \rho_0$; where ρ_0 is a constant independent of x_0.

3) If u is bounded then obviously $u \in L^{p,\lambda}(\Omega)$ (cf. also the proposition below).

4) The embedding $L^{p,\lambda}(\Omega) \hookrightarrow L^p(\Omega)$ is a continuous operator (by Hölder's inequality).

5) $L^{p,0}(\Omega) \cong L^p(\Omega)$.

6) For $\frac{n-\lambda}{p} \ge \frac{n-\mu}{q}$; $p \le q$ we have $L^{q,\mu} \subset L^{p,\lambda}$:

$$\begin{aligned}\int_{\Omega(x_0,\rho)} |u|^p dx &\le \left(\int_{\Omega(x_0,\rho)} |u|^q dx\right)^{p/q} |\Omega(x_0,\rho)|^{1-p/q} \\ &\le \left\{||u||^q_{L^{q},\mu}\rho^\mu\right\}^{p/q} \rho^{n(1-p/q)} \\ &\le ||u||^q_{L^{q,\mu}} \rho^{\overbrace{\mu p/q+n-n\,p/q}^{\ge 1}} .\end{aligned}$$

7) $u \in L^{p,\lambda}(\Omega)$ does *not* imply that $u \in L^{q,\mu}(\Omega)$ for $q > p$!

Proposition 3.1 $L^{p,n}(\Omega) \cong L^\infty(\Omega)$.

PROOF: If $u \in L^\infty(\Omega)$, then

$$\int_{\Omega(x_0,\rho)} |u|^p dx \le c|B_1|\ ||u||^p_\infty \rho^n \qquad (\ |B_1| \text{ is the volume of the unit ball in } \mathbb{R}^n)$$

i.e. $u \in L^{p,n}(\Omega)$. If on the other hand $u \in L^{p,n}(\Omega)$ then for a.e. $x_0 \in \Omega$

$$u(x_0) = \lim_{\rho\to 0} \frac{1}{|\Omega(x_0,\rho)|} \int_{\Omega(x_0,\rho)} u dx$$

(Lebesgue-theorem, see next chapter!).

So we have $|u(x_0)| \le \lim_{\rho\to 0} \fint_{\Omega(x_0,\rho)} |u| dx \le \lim_{\rho\to 0} \left(\fint_{\Omega(x_0,\rho)} |u|^p dx\right)^{1/p}$ and by the assumption the right hand side is equibounded by a constant independent of x_0, thus $u \in L^\infty(\Omega)$.

Remark: $L^{p,\lambda}(\Omega) = \{0\}$ for $\lambda > n$.

PROOF: By Lebesgue's theorem we have

$$|u(x_0)| \leq \frac{1}{|\Omega(x_0,\rho)|^{1/p}} \left(\int_{\Omega(x_0,\rho)} |u|^p dx \right)^{1/p} \leq c\rho^{-n/p}\rho^{\lambda/p} = c\rho^{\lambda - n/p}$$

as $\lambda - n > 0$, $\rho \to 0$ implies $u(x_0) = 0$ and as x_0 was arbitrary $u = 0$.

We now come to the definition of the *Campanato-spaces* $\mathcal{L}^{p,\lambda}$:

$$\mathcal{L}^{p,\lambda}(\Omega) := \left\{ u \in L^p(\Omega) \middle| \sup_{\substack{x_0 \in \Omega \\ 0<\rho<\operatorname{diam}\Omega}} \rho^{-\lambda} \int_{\Omega(x_0,\rho)} |u - u_{x_0,\rho}|^p dx < \infty \right\}$$

where $u_{x_0,\rho} := \fint_{\Omega(x_0,\rho)} u dx$.

Remarks:

1) Set $[u]^p_{p,\lambda} := \sup\limits_{\substack{x_0 \in \Omega \\ 0<\rho<\operatorname{diam}\Omega}} \rho^{-\lambda} \int_{\Omega(x_0,\rho)} |u - u_{x_0,\rho}|^p dx$ then $[\ ]^p_{p,\lambda}$ is a seminorm ($[u]_{p,\lambda} = 0$ iff $u =$ constant). With the norm $\|\ \|_{\mathcal{L}^{p,\lambda}(\Omega)} := \|\ \|_{L^p(\Omega)} + [\ ]_{p,\lambda}$ $\mathcal{L}^{p,\lambda}(\Omega)$ becomes a Banach-space.

2) If $u \in C^{0,\alpha}(\bar{\Omega})$ then $\int_{\Omega(x_0,\rho)} |u - u_{x_0,\rho}|^2 dx \leq \rho^{n+2\alpha}$ for all ρ with $0 < \rho <$ diam Ω, i.e. $C^{0,\alpha}(\bar{\Omega}) \subset \mathcal{L}^{2,\lambda}(\Omega)$ for $\lambda = n + 2\alpha$ (we shall actually show that the two spaces are isomorphic cf. p. 42).

3) For $\frac{n-\lambda}{p} \geq \frac{n-\mu}{q}$, $p \leq q$ we have $\mathcal{L}^{q,\mu} \subset \mathcal{L}^{p,\lambda}$.

Proposition 3.2 *For* $0 \leq \lambda \leq n$ *we have* $L^{p,\lambda}(\Omega) \cong \mathcal{L}^{p,\lambda}(\Omega)$.

PROOF: Suppose $u \in L^{p,\lambda}(\Omega)$ then

$$\rho^{-\lambda} \int_{\Omega(x_0,\rho)} |u - u_{x_0,\rho}|^p dx \leq 2^{p-1} \left\{ \rho^{-\lambda} \int_{\Omega(x_0,\rho)} |u|^p dx + \rho^{-\lambda} \underbrace{|u_{x_0,\rho}|^p}_{\leq \fint_{\Omega(x_0,\rho)} |u|^p} |\Omega(x_0,\rho)| \right\} < \infty$$

hence $u \in \mathcal{L}^{p,\lambda}$.

Now assume $u \in \mathcal{L}^{p,\lambda}(\Omega)$. We find

$$\begin{aligned} \rho^{-\lambda} \int\limits_{\Omega(x_0,\rho)} |u|^p dx &\leq 2^{p-1} \left\{ \rho^{-\lambda} \int\limits_{\Omega(x_0,\rho)} |u - u_{x_0,\rho}|^p dx + \rho^{-\lambda} |u_{x_0,\rho}|^p |\Omega(x_0,\rho)| \right\} \\ &\leq 2^{p-1} \left\{ \rho^{-\lambda} \int\limits_{\Omega(x_0,\rho)} |u - u_{x_0,\rho}|^p dx + c\rho^{n-\lambda} |u_{x_0,\rho}|^p \right\} . \end{aligned} \tag{3.1}$$

We have to estimate $|u_{x_0,\rho}|$: Take $0 < r < R$ then

$$|u_{x_0,\rho}|^p \leq 2^{p-1} \left\{ |u_{x_0,R}|^p + |u_{x_0,R} - u_{x_0,\rho}|^p \right\} . \tag{3.2}$$

In order to estimate the second term on the right hand side we note that

$$\begin{aligned} &\mathrm{r}^n |u_{x_0,R} - u_{x_0,r}|^p \\ &\leq |u_{x_0,R} - u_{x_0,r}|^p |\Omega(x_0,r)| \\ &\leq \int\limits_{\Omega(x_0,r)} |u_{x_0,R} - u_{x_0,r}|^p dx \\ &\leq 2^{p-1} \left\{ \int\limits_{\Omega(x_0,R)} |u(x) - u_{x_0,R}|^p dx + \int\limits_{\Omega(x_0,r)} |u(x) - u_{x_0,r}|^p dx \right\} \\ &\leq 2^{p-1} b \cdot [R^\lambda + r^\lambda] [u]_{p,\lambda}^p \leq c R^\lambda [u]_{p,\lambda}^p . \end{aligned}$$

We divide by r^n and take then the p-th root to get:

$$|u_{x_0,R} - u_{x_0,r}| \leq \tilde{c} [u]_{p,\lambda} R^{\lambda/p} r^{-n/p} .$$

Set $R_i := R2^{-i}$ then

$$|u_{x_0,R_i} - u_{x_0,R_{i+1}}| \leq \tilde{c} [u]_{p,\lambda} 2^{i(n-\lambda)/p} R^{\lambda - n/p} .$$

$$\begin{aligned} |u_{x_0,R} - u_{x_0}, R_{h+1}| &\leq \sum_{i=0}^{h} |u_{x_0,R_i} - u_{x_0,R_{i+1}}| \\ &\leq \tilde{C} [u]_{p,\lambda} R^{\lambda - n/p} \sum_{i=0}^{h} 2^{i(n-\lambda)/p} \\ &\leq d [u]_{p,\lambda} R_{h+1}^{\lambda - n/p} . \end{aligned} \tag{3.3}$$

Choose R (with diam $\Omega < R \leq 2$ diam Ω) in such a way that for some h $\rho = R_{h+1}$. Insert (3.3) in (3.2) and (3.2) in (3.1) to find

$$\rho^{-\lambda} \int\limits_{\Omega(x_0,\rho)} |u|^p dx \leq 2^{p-1} \left\{ \rho^{-\lambda} \int\limits_{\Omega(x_0,\rho)} |u - u_{x_0,\rho}|^p dx \right.$$

$$\left. + c\rho^{n-\lambda} 2^{p-1} \left| \;\fint\limits_{\Omega(x_0,\rho)} u dx \right|^p + c d [u]_{p,\lambda} 2^{p-1} \right\} < \infty$$

thus $u \in L^{p,\lambda}(\Omega)$.

Remarks:

1) $L^\infty(\Omega) \subsetneqq \mathcal{L}^{p,n}(\Omega)$ e.g. look at $\log x \in \mathcal{L}^{1,1}(]0,1[)$.

2) For Ω a cube in $\mathbb{R}^n$ $\mathcal{L}^{p,n}(\Omega)$ is isomorphic to the space of functions of bounded mean oscillation (= John-Nirenberg-space) (see the next chapter for more details).

3) With the construction in the above proof one finds for $k < h$:

(∗) $$|u_{x_0,R_k} - u_{x_0,R_h}| \leq c[u]_{p,\lambda} R_k^{\lambda - n/p} .$$

As we have seen Campanato-spaces can be identified with Morrey-spaces in the intervall $0 \leq \lambda < n$. For $\lambda > n$ we have the following

Theorem 3.1 (Campanato, [7]) *For $n < \lambda \leq n + p$ holds*

$$\mathcal{L}^{p,\lambda}(\Omega) \cong C^{0,\alpha}(\bar{\Omega}) \qquad \text{with } \alpha = \frac{\lambda - n}{p}$$

whereas for $\lambda > n + p$

$$\mathcal{L}^{p,\lambda}(\Omega) = \{constants\}.$$

PROOF: From remark 3) p. 41 and the assumption that $\lambda > n$ we conclude that u_{x_0,R_k} is a Cauchy-sequence for all $x_0 \in \Omega$. Thus by Lebesgue's theorem

$$u_{x_0,R_k} \longrightarrow \tilde{u}(x_0) \qquad (k \to \infty)$$

also if $r < R$

$$u_{x_0,R_k} \longrightarrow \tilde{u}(x)$$

and $\tilde{u}$ is a representative of u.

Now let $h \to \infty$ in $(*)$ then

$$|u_{x_o,R_k} - u(x_0)| \le c[u]_{p,\lambda} R_k^{\frac{\lambda-n}{p}}$$

and we conclude that the above convergence is actually uniform and that u is continuous.

We finally show that u is Hölder-continuous:
Take $x, y \subset \Omega$ and set $R = |x-y|$, $\alpha = \frac{\lambda-n}{p}$.

$$|u(x) - u(y)| \le \Bigg|\underbrace{u_{x,2R} - u(x)}_{\le cR^\alpha}\Bigg| + |u_{x,2R} - u_{y,2R}| + \Bigg|\underbrace{u_{y,2R} - u(y)}_{\le cR^\alpha}\Bigg|$$

and

$$\begin{aligned}
& |\Omega(x,2R) \cap \Omega(y,2R)|\ |u_{x,2R} - u_{y,2R}| \\
\le\ & \int\limits_{\Omega(x,2R)} \Big|u(z) - u_{x,2R}\Big| dz + \int\limits_{\Omega(y,2R)} \Big|u(z) - u_{y,2R}\Big| dz \\
\le\ & \left(\int\limits_{\Omega(x,2R)} |u(z) - u_{x,2R}|^p \right)^{1/p} |\Omega(x,2R)|^{1-1/p} \\
& + \left(\int\limits_{\Omega(y,2R)} |u(z) - u_{y,2R}|^p \right)^{1/p} |\Omega(y,2R)|^{1-1/p} \\
\le\ & 2[u]_{p,\lambda} R^{\lambda/p} R^{n(1-1/p)} \\
=\ & c[u]_{p,\lambda} R^\alpha
\end{aligned}$$

Now as $R = |x-y|$ we have shown:

$$[u]_{C^{0,\alpha}(\bar{\Omega})} = \sup_{\substack{x,y\in\Omega \\ x\neq y}} \frac{|u(x) - u(y)|}{|x-y|} \le c[u]_{\mathcal{L}^{p,\lambda}(\Omega)} \ .$$

Actually, we have

$$\sup_{x\in\Omega} |u(x)| + [u]_{C^{0,\alpha}(\bar{\Omega})} \le c\left(||u||_{L^p(\Omega)} + [u]_{\mathcal{L}^{p,\lambda}(\Omega)}\right) \ .$$

This follows from the following estimate:

$$|u(x)| \le |u(y)| + |u(y) - u(x)| \le c \left(\int\limits_\Omega |u|^p \right)^{1/p} + d[u]_{p,\lambda} (\mathrm{diam}\Omega)^\alpha$$

Consequences

1) If we have $\int\limits_{\Omega(x_0,\rho)} |u - u_{x_0,\rho}|^p dx \leq c\rho^\lambda$ for all $\rho \leq \rho_0$ (instead of $\rho <$ diamΩ), the result remains true and we have a local version of the above theorem.

2) Suppose that

$$\int\limits_{B_\rho(x_0)} |u - u_{x_0,\rho}|^2 dx \leq c\rho^\lambda \quad \text{for all } x_0 \in \Omega \text{ and } \rho < \text{dist }(x_0, \partial\Omega).$$

Then for $\lambda > n$ we conclude that u is Hölder-continuous in every subdomain of Ω.

3) **Morrey's theorem on the growth of the Dirichlet-integral.** *Let u be in $H^{1,p}_{\text{loc}}(B_1)$ and suppose that*

$$\int\limits_{B_\rho(x)} |Du|^p dx \leq \rho^{n-p+p\alpha} \qquad \textit{for all } \rho < \textit{dist }(x, B_1)$$

then $u \in C^{0,\alpha}_{\text{loc}}(B_1)$.

PROOF: We use Poincarés inequality:

$$\int\limits_{B_\rho(x_0)} |u - u_{x_0,\rho}|^p dx \leq c\rho^p \int\limits_{B_\rho(x_0)} |Du|^p dx \leq c\rho^{n+p\alpha}$$

i.e. $u \in \mathcal{L}^{p,\lambda}$ and therefore by 1) $u \in C^{0,\alpha}(B_1)$.

4) Global version of 3):

Let Ω be a Lipschitz-domain and $p > n$. For all ρ less that some ρ_0 and a $u \in H^{1,p}(\Omega)$ we have

$$\int\limits_{\Omega(x_0,\rho)} |u - u_{x_0,\rho}|^p dx \leq c\rho^p \int\limits_{\Omega(x_0,\rho)} |Du|^p dx \ ,$$

where c depends only on the geometry of Ω; then we find

$$\int\limits_{\Omega(x_0,\rho)} |Du| dx \leq \left(\int\limits_{\Omega(x_0,\rho)} |Du|^p dx \right)^{1/p} |\Omega(x_0,\rho)|^{1-1/p} \leq c||u||_{H^{1,p}(\Omega)} \rho^{n-n/p} \, .$$

Again by Poincaré's inequality:

$$\int\limits_{\Omega(x_0,\rho)} |u - u_{x_0,\rho}| dx \leq \rho \int\limits_{\Omega(x_0,\rho)} |Du| dx \leq c\rho^{n+1-n/p} ||u||_{H^{1,p}(\Omega)}$$

and we have proved: *if* $u \in H^{1,p}(\Omega)$, $p > n$, *then* $u \in \mathcal{L}^{1,n+(1-n/p)}(\Omega)$, i.e. $u \in C^{0,1-n/p}(\Omega)$.

A consequence of this argument is Morrey's theorem (or Sobolev's theorem): For $p > n$ the embedding

$$H^{1,p}(\Omega) \hookrightarrow C^{0,1-n/p}(\bar{\Omega})$$

is a continuous operator.

5)

$$\begin{array}{rcl} Du \in L^{p,\lambda}(\Omega) & \Rightarrow & u \in \mathcal{L}^{p,\lambda+p}(\Omega) \\ (Du \in L^{2,n-2+2\alpha} & \Rightarrow & u \in \mathcal{L}^{2,n+2\alpha}(\Omega)) \ . \end{array}$$

3.2 (Interior-)Schauder-estimates for elliptic systems in divergence form

A useful lemma. We look at a function $\Phi(\rho)$ which is nonnegative and nondecreasing.

Suppose

(+) $\quad \Phi(\rho) \leq A\left[(\rho/R)^\alpha + \varepsilon\right]\Phi(R) + BR^\beta$ *for all* $\rho < R \leq R_0$,

where A, B, α *and* β *are nonnegative constants and* $\alpha > \beta$. *Then there exists* $\varepsilon_0 = \varepsilon_0(\alpha, \beta, A)$ *such that the following holds:*
If (+) is true for some $\varepsilon < \varepsilon_0$ *then*

$$\Phi(\rho) \leq c(\alpha, \beta, A)\left[(\rho/R)^\alpha \Phi(R) + B\rho^\beta\right]$$

(roughly speaking: one can replace R *by* ρ *in the "B-term" and omit* ε*.)*

PROOF: Set $\rho = \tau R$ for $0 < \tau < 1$ then

$$\Phi(\tau R) \leq \tau^\alpha A[1 + \varepsilon\tau^{-\alpha}]\Phi(R) + BR^\beta \ .$$

Fix $\gamma \in]\beta, \alpha[$ and choose τ such that $A\tau^\alpha = \tau^\gamma$ and ε_0 such that $\varepsilon_0\tau^{-\alpha} < 1$. Our assumption now becomes:

$$\Phi(\tau R) \leq \tau^\gamma \Phi(R) + BR^\beta$$

and hence

$$\begin{array}{rcl} \Phi(\tau^{k+1}R) & \leq & \tau^\gamma \Phi(\tau^k R)^\beta + B(\tau^k R)^\beta \\ & \leq & \tau^\gamma\left[\tau^\gamma \Phi(\tau^{k-1}R) + B(\tau^{k-1}R)^\beta\right] + B\tau^{\beta k}R^\beta \end{array}$$

or

$$\Phi(\tau^{k+1}R) \leq \tau^{(k+1)\gamma}\Phi(R) + B\tau^{\beta k}R^\beta \sum_{j=0}^{k} \tau^{j(\gamma-\beta)} \ .$$

Because $\rho < R$ there exists k with $\tau^{k+1}R \leq \rho < \tau^k R$ and

$$\Phi(\rho) \leq \Phi(\tau^{k+1}R) \leq \tau^{(k+1)\gamma}\Phi(R) + cB(\tau^k R)^\beta < (\rho/R)^\gamma \Phi(R) + cB\tau^{-\beta}\rho^\beta \,.$$

Let us consider a solution $u \in H^{1,k}_{\text{loc}}(\Omega, \mathbb{R}^N)$ of

$$(*) \qquad D_\alpha(A^{\alpha\beta}_{ij} D_\beta u^j) = 0$$

where we suppose that $A^{\alpha\beta}_{ij}$ = const. and satisfy a $L - H$-condition.

Theorem 3.2 *For all* $\rho < R \leq R_0$ *we have:*

$$\int\limits_{B_\rho(x_0)} |u|^2 dx \leq c(\rho/R)^n \int\limits_{B_R(x_0)} |u| dx \tag{3.4}$$

$$\int\limits_{B_\rho(x_0)} |u - u_{x_0,\rho}|^2 dx \leq c(\rho/R)^{n+2} \int\limits_{B_R(x_0)} |u - u_{x_0,R}|^2 dx \tag{3.5}$$

where $u_{x_0,\rho} = \fint\limits_{B_\rho(x_0)} u dx$.

PROOF:

(1) Since u is a solution of $(*)$ by Hilbert-space regularity, we have

$$||u||_{H^{k,2}(B_{R/2})} \leq c(k,R)||u||_{L^2(B_R)} \qquad \text{cf. p. 31).}$$

Thus for $\rho < R/2$ we find

$$\begin{aligned} \int\limits_{B_\rho(x_0)} |u|^2 dx &\leq c(n)\rho^n \sup_{B_\rho(x_0)} |u|^2 \\ &\leq c(n)\rho^n ||u||^2_{H^{k,2}(B_{R/2}(x_0))} \\ &\leq c(k,R)\rho^n \int\limits_{B_R(x_0)} |u|^2 dx \end{aligned}$$

(observe that for the second estimate we have used Sobolev's embedding theorem!), moreover by a rescaling argument: $c(k,R) = c(k)R^{-n}$.

(2) is a consequence of (1) (we use in addition Poincaré- and Caccioppoli-in-

equality):

$$\begin{aligned}\int\limits_{B_\rho(x_0)} |u-u_{\rho,x_0}|^2dx &\leq c\rho^2 \int\limits_{B_\rho(x_0)} |Du|^2dx\\ &\leq \tilde{c}\rho^2(\rho/R)^n \int\limits_{B_{R/2}(x_0)} |Du|^2dx\\ &= \tilde{c}(\rho/R)^{n+2}R^2 \int\limits_{B_{R/2}(x_0)} |Du|^2dx\\ &\leq \tilde{c}(\rho/R)^{n+2} \int\limits_{B_R(x_0)} |u-u_{x_0,R}|^2dx \ .\end{aligned}$$

Remarks:

1) (1) and (2) hold for all derivatives $D^\gamma u$ because for a system with constant coefficients any derivative of a solution is again a solution.

2) One knows that, if u is a subharmonic function, then

$$R \longmapsto \frac{1}{R^n}\int\limits_{B_R} udx$$

is nondecreasing. (1) is in some way a generalization of this: If u is harmonic $|u|^2$ and $|Du|^2$ are subharmonic.

3) There exists a unique polynomial p_{m-1} such that

$$\int\limits_{B_\rho} D^\gamma(u-p_{m-1})dx = 0$$

for all γ with $|\gamma| \leq m-1$ and the following Poincaré-inequality holds:

$$\int\limits_{B_\rho} |u-p_{m-1}|^2dx \leq c\rho^{2m}\int\limits_{B_\rho} |D^m u|^2dx \ .$$

As in the proof of the theorem above we then get

$$\int\limits_{B_\rho} |u-p_{m-1}|^2dx \leq (\rho/R)^{n+2m}\int\limits_{B_R} |u|^2dx \ .$$

If we assume that u is a solution of $(*)$ (p. 45) in $\mathbb{R}^n$ such that for large x $|u(x)| \leq A|x|^\sigma$, $\sigma = [m]+1$ then by letting R go to ∞ we get

$$\int\limits_{B_\rho} |u-p_{m-1}|^2dx = 0 \ .$$

Thus:

If the growth of u is polynomial, then u actually is a polynomial.

We now look at *nonhomogeneous elliptic systems with constant coefficients.*

Suppose $u \in H^1_{\text{loc}}(\Omega)$ is a solution of

$$D_\alpha(A^{\alpha\beta}_{ij} D_\beta u^j) = -D_\alpha f^i_\alpha\,, \qquad i = 1, \ldots, N$$

where the $A^{\alpha\beta}$ are constant and satisfy a $L-H$-condition.

Theorem 3.3 *Suppose $f \in \mathcal{L}^{2,\lambda}_{\text{loc}}(\Omega)$ then $Du \in \mathcal{L}^{2,\lambda}_{\text{loc}}(\Omega)$ for $0 \le \lambda < n+2$.*

Corollary 3.1 *If $f \in C^{0,\mu}_{\text{loc}}$ then $Du \in C^{0,\mu}_{\text{loc}}(\Omega)$. ($\mu \in (0,1)$ cf. Campanato's theorem p. 41).*

Proof of the theorem: We split u in $B_R \subset \Omega := v + (u-v) = v + w$ where v is the solution (which exists by Lax-Milgram) of the Dirichlet-boundary-value-problem:

$$\begin{cases} D_\alpha(A^{\alpha\beta} D_\beta v^j) = 0 & i = 1, \ldots, N \quad \text{in } B_R \\ v = u & \text{on } \partial B_R\,. \end{cases}$$

Hence by the above (Dv is also a local solution): for all $\rho < R$

$$\int_{B_\rho} |Dv - (Dv)_{x_0,\rho}|^2 dx \le c(\rho/R)^{n+2} \int_{B_R} |Dv - (Dv)_{x_0,\rho}|^2 dx\;.$$

Now we have

$$\begin{aligned} &\int_{B_\rho} |Du - (Du)_{x_0,\rho}|^2\, dx \\ &\quad = \int_{B_\rho} |Dv + Dw - (Dv)_{x_0,\rho} - (Dw)_{x_0,\rho}|^2 dx \\ &\quad \le c(\rho/R)^{n+2} \int_{B_R} |Dv - (Dv)_{x_0,\rho}|^2 dx + \int_{B_\rho} |Dw - (Dw)_{x_0,\rho}|^2 dx \end{aligned}$$

replace $v = w - u$:

$$\int_{B_\rho} |Du - (Du)_{x_0,\rho}|^2\, dx \le c(\rho/R)^{n+2} \int_{B_R} |Du - (Du)_{x_0,\rho}|^2\, dx + \tilde{c} \int_{B_r} |D(u-v)|^2 dx$$

and because of the $L-H$-condition

$$\int_{B_R} |D(u-v)|^2 dx \le \int_{B_R} AD(u-v)D(u-v)dx = \int_{B_R} (f-f_{x_0,R})D(u-v)dx$$

we see that $\int_{B_R} |D(u-v)|^2 dx$ is estimated by $\int_{B_R} |f-f_{x_0,R}|^2 dx$. We conclude therefore:

$$\int_{B_\rho} |Du-(Du)_{x_0,\rho}|^2 dx \le c(\rho/R)^{n+2} \int_{B_R} \left|Du-(Du)_{x_0,R}\right|^2 dx + c[f]^2_{\mathcal{L}^{2,\lambda}(\Omega)} R^\lambda$$

now by the lemma p. 39 R^λ in the second term can be replaced by ρ^λ and the theorem is proved.

Actually we have proved even more, namely:

$$[Du]_{\mathcal{L}^{2,\lambda}(\tilde\Omega)} \le c(\Omega,\tilde\Omega)\left[\|u\|_{L^2(\Omega)} + [f]_{\mathcal{L}^{2,\lambda}(\Omega)}\right] \qquad \text{with } \tilde\Omega \subset\subset \Omega \ .$$

All this holds for constant coefficients. We now prove a similar result for *variable coefficients:*

Theorem 3.4 *Suppose:* $A^{\alpha\beta}_{ij} \in C^0(\Omega)$ *and satisfy a* $L-H$*-condition. If* $f^i_\alpha \in L^{2\lambda}(\Omega)$ $(\cong \mathcal{L}^{2,\lambda}(\Omega)$ *for* $0 \le \lambda \le n)$*, then* $Du \in L^{2,\lambda}_{\text{loc}}(\Omega)$.

PROOF:

$$D_\alpha\left(A^{\alpha\beta}_{ij}(x_0)D_\beta u^j\right) = D_\alpha\left\{\left(A^{\alpha\beta}_{ij}(x_0)-A^{\alpha\beta}_{ij}(x)\right)D_\beta u^j + f^i_\alpha\right\} =: D_\alpha\{F^i_\alpha\} \ .$$

As before

$$\begin{aligned}\int_{B_\rho} |Du|^2 dx &\le c(\rho/R)^n \int_{B_R} |Du|^2 dx + c\int_{B_R} |D(u-v)|^2 dx \\ &\le c(\rho/R)^n \int_{B_R} |Du|^2 dx + c\int_{B_R} |f|^2 dx + c\omega(R)\int_{B_R} |Du|^2 dx\end{aligned}$$

Now $\int_{B_R} |f|^2 dx \le R^\lambda$ as $f \in L^{2,\lambda}(\Omega)$ and if we choose R_0 sufficiently small and $\rho < R < R_0$, then $\omega(R)$ is small and again we can apply the lemma which gives the result

Theorem 3.5 (Hölder-continuous case) *Let* u *be a solution of*

$$D_\alpha\left(A^{\alpha\beta}_{ij}(x)D_\beta u^j\right) = D_\alpha f^i_\alpha \qquad j=1,\dots,N$$

with $A^{\alpha\beta}_{ij} \in C^{0,\mu}$ *and satisfy a* $L-H$*-condition.*

If $f^i_\alpha \in C^{0,\mu}$ *then* $Du \in C^{0,\mu}$ *and moreover*

$$[Du]_{C^{0,\mu}(\tilde\Omega)} \le c\left\{\|u\|_{L^2(\Omega)} + [f]_{C^{0,\mu}(\tilde\Omega)}\right\}$$

PROOF:

$$\int_{B_\rho} |Du-(Du)_\rho|^2\,dx \le c((\rho/R)^{n+2}\int_{B_R} |Du-(Du)_{x_0,r}|^2\,dx + \underbrace{\int_{B_R} |f-f_R|^2 dx}_{\le cR^{n+2\mu}} + \underbrace{\omega(R)}_{\le cR^{2\mu}} \int_{B_R} \underbrace{|Du|^2 dx}_{\le R^{n-\varepsilon}c(\varepsilon)}$$

The number corresponding to β in the lemma is $n+2\mu-\varepsilon$. We conclude that Du is Hölder-continuous (but actually not with the exact exponent) so it is bounded and therefore $\int_{B_R} |Du|^2 dx \le a \int_{B_R} dx \le \tilde{a}R^n$, i.e. the exponent is in fact $n+2\mu$ and the theorem is proved.

3.3 (Interior)-Schauder estimates for elliptic systems in non-divergence form

In this section we are concerned with systems in non-divergence form. If A is sufficiently smooth, there is of course no difference between divergence- and non-divergence-form. But here we shall have only continuity in general:

$$A_{ij}^{\alpha\beta} u^j_{x_\alpha x_\beta} = f^i \qquad i=1,\ldots,N \ .$$

For $u \in H^2_{\text{loc}}$ and $A_{ij}^{\alpha\beta} \in C^{0,\mu}$ we have

Theorem 3.6 *If $f^i \in C^{0,\mu}$ then $u_{x_\alpha x_\beta} \in C^{0,\mu}$ and moreover*

$$[u_{x_\alpha x_\beta}]_{C^{0,\mu}} \le c\left[||u_{x_\alpha x_\beta}||_{L^2} + [f]_{C^{0,\mu}}\right] \ .$$

PROOF: *Step 1:* Suppose $A=$ const and $f=0$, then we are in a variational situation and therefore

$$\int_{B_\rho} |u_{xx}-(u_{xx})_\rho|^2 dx \le c(\rho/R)^{n+2} \int_{B_R} |u_{xx}-(u_{xx})_R|^2\,dx \ .$$

Step 2: Suppose $A=$ const and $f \ne 0$.
Set $z^i(x) := u^i(x) - 1/2(u^i_{x_\alpha x_\beta})_R x_\alpha x_\beta$ thus $z^i_{x_\gamma x_\sigma} = u^i_{x_\gamma x_\sigma} - (u^i_{x_\gamma x_\sigma})_R$ and we have

$$A_{ij}^{\alpha\beta} z^j_{x_\alpha x_\beta} = f^i - f^i_R \ .$$

We now split $z = v + (z-v)$ where v is a solution of

$$\begin{cases} Av_{xx} &= 0 \quad \text{on} \quad B_R \\ v &= z \quad \text{on} \quad \partial B_R \end{cases}$$

and clearly

$$\begin{cases} A(z-v)_{xx} &= f - f_R \quad \text{on} \quad B_R \\ z - v &= 0 \quad\quad\;\; \text{on} \quad \partial B_R \end{cases}$$

As in the variational case then, taking into account the estimate (p. 48)

$$\int_{B_R} \left|D^2(z-v)\right|^2 dx \le \int_{B_R} |f - f_R|^2 dx ,$$

we get

$$\int_{B_\rho} \left|z_{xx} - (z_{xx})_\rho\right|^2 dx \le c(\rho/R)^{n+2} \int_{B_R} \left|z_{xx} - (z_{xx})_R\right|^2 dx + c \int_{B_R} |f - f_R|^2 dx$$

i.e.

$$\int_{B_\rho} \left|u_{xx} - (u_{xx})_\rho\right|^2 dx \le c(\rho/R)^{n+2} \int_{B_R} \left|u_{xx} - (u_{xx})_R\right|^2 dx + c \int_{B_R} |f - f_R|^2 dx .$$

The proof now continues as in the variational case by writing the equation

$$A(x) u_{xx} = f$$

as $(*)$ $A(x_0)u_{xx} = [A(x_0) - A(x)]\, u_{xx} + f$ and by using step 2 with f substituted by the right hand side of $(*)$.

3.4 Regularity up to the boundary: Dirichlet-boundary-value problem

In this section we consider solutions of the Dirichlet-boundary-value problem

$$D_\alpha \left(A_{ij}^{\lambda\beta}(x) D_\beta u^j\right) = D_\alpha f_\alpha^i \quad \text{in } \Omega \ , \qquad u = g \quad \text{on } \partial\Omega \in C^{1,\mu}$$

with $f_\alpha^i \in C^{0,\mu}(\Omega)$ and $g \in C^{1,\mu}(\bar{\Omega}, \mathbb{R}^N)$ and we shall extend the local regularity result of section 3.2 we shall only point out the new aspects and just sketch the rest.

First we remark that we can reduce to zero boundary values, if we look at $U = u - g$.

Our assumption on the boundary (i.e. $C^{1,\mu}$) says that (locally) there exists a diffeomorphism

$$G : U \to B_R^+ \ ,$$

which flattens the boundary i.e. maps $\partial\Omega \cap U$ onto Γ_R:

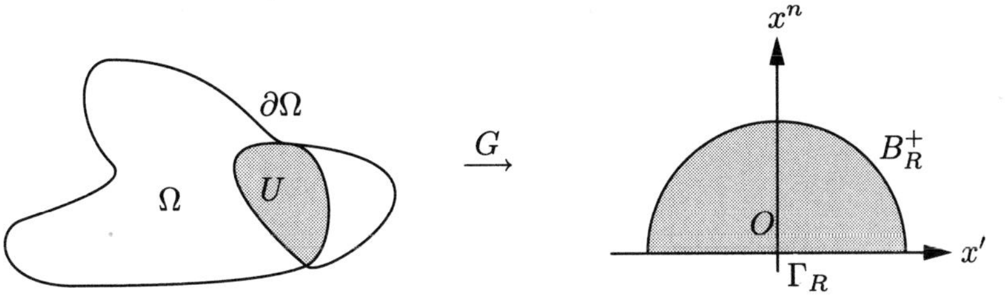

Now clearly it suffices first to find estimates on small balls. Then one gets global estimates on Ω by using a covering-argument.

So we may suppose that $\Omega = B_R^+$ and $\partial\Omega = \Gamma_R$. We are looking for estimates corresponding to (3.4), (3.5) in 3.2 (cf. p.45):

Step 1: Suppose $A_{ij}^{\alpha\beta}$ = const. and $f_\alpha^i = 0$. For $\rho < R/2$ we have

$$\begin{aligned}\int_{B_\rho^+} |Du|^2 dx &\leq c\rho^n \sup_{B_\rho^+} |Du|^2 \\ &\leq c(k)\rho^n ||u||_{H^k(B_{R/2}^+)} \\ &\leq d(R)\rho^n \int_{B_R^+} |u|^2 dx \\ &\leq c(R)\rho^n \int_{B_R^+} |u_{x_n}|^2 dx\end{aligned}$$

this corresponds to (3.4) (the second estimate is Sobolev's embedding theorem!).

The equivalent to (3.5) is the following:

$$\int_{B_\rho^+} |Du - (Du)_\rho|^2 dx \leq c\rho^2 \int_{B_\rho^+} |D^2u|^2 dx \leq c\rho^{n+2} \int_{B_R^+} |u_{x_n}|^2 dx \ .$$

The same calculation can be carried out for $u - \lambda x_n$ (which is also a solution with zero boundary value) then in the above estimated we have to replace u_{x_n} by $u_{x_n} - \lambda$ (λ = constant).

Step 2: There are essentially three possible cases for balls centered at x_0 inside B_R^+:

(a) $\text{dist}(x_0, \Gamma) > R/2$:
in this case for $\rho < R/2$, we have

$$\int_{B_\rho} |u - u_{x_0,\rho}|^2 dx \leq c(\rho/R/2))^{n+2} \int_{B_{R/2}} |u - u_{x_0,R/2}|^2 dx$$

(b) dist $(x_0, \Gamma) = r \leq R/2$, $B_\rho(x_0) \subset B^+$

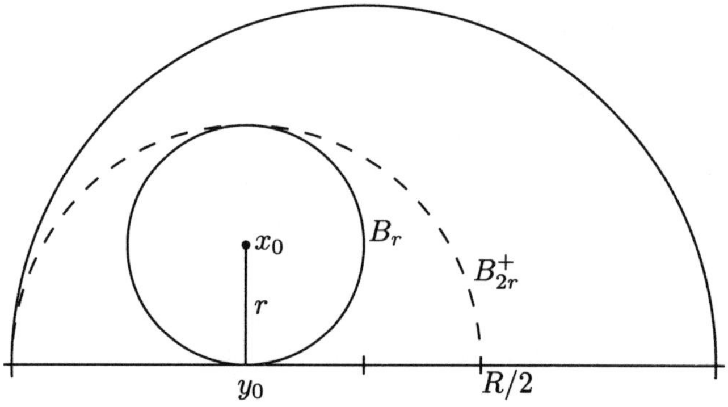

here we first estimate $I(B_\rho)$ (:= the integral over B_ρ) by $I(B_r)$, then $I(B_r)$ by $I(B^+_{2r})$ and finally $I(B^+_{2r})$ by $I(B_{R(2)})$:

$$\begin{aligned} \int_{B_\rho} |u - u_{x_0,\rho}|^2 dx &\leq c(\rho/r)^{n+2} \int_{B_r(x_0)} |u - u_{x_0,r}|^2 dx \\ &\leq c(\rho/r)^{n+2} \int_{B^+_{2r(y_0)}} |u - u_{x_0,2r}|^2 dx \\ &\leq \tilde{c}(\rho/r)^{n+2}(r/R/2)^{n+2} \int_{B_{R/2}} |u - u_{x_0,R/2}|^2 dx \end{aligned}$$

(c) $\operatorname{dist}(x_0, \Gamma) = r < R/2, B_\rho(x_0) \not\subset B^+$
here we first estimate $I(B_\rho \cap B^+)$ by $I\left(B^+_{2\rho}(y_0)\right)$ and finally $I\left(B^+_{2\rho}(y_0)\right)$ by $I(B_{R/2})$.

The nonvariational case

We look at a solution of

$$\begin{cases} A^{\alpha\beta}(x) u^j_{x_\alpha x_\beta} &= f^i \in C^{2,\mu} \quad \text{in} \quad \Omega \\ u &= g \in C^{2,\mu} \quad \text{on} \quad \partial\Omega \in C^{2,\mu} \end{cases}$$

As in 3.2 we find the equivalent of (3.4):

$$\int_{B^+_\rho} |u_{xx}|^2 dx \leq c(\rho/R)^n \int_{B^+_R} |u_{xx}|^2 dx$$

$$(*) \qquad \int_{B^+_\rho} \left|u_{xx} - (u_{xx})_\rho\right|^2 dx \leq c(\rho/R)^{n+2} \int_{B_R} |u_{xx} - (u_{xx})_\rho|^2 dx$$

moreover we have

$$||u||_{C^{2,\mu}(\bar{\Omega})} \leq c\left\{||f||_{C^{0,\mu}(\bar{\Omega})} + ||u_{xx}||_{L^2(\Omega)}\right\} .$$

The proof is the same as in 3.2 p. 49 with the only exception that in the second step instead of $z = u - 1/2(u_{x_\alpha x_\beta})_R x_\alpha x_\beta$ one takes

$$z = u - 1/2(a_{ij}^{nn})^{-1}(f^j)_R x_\beta^2 ,$$

then there appears a term $\int\limits_{B_R} (f - f_R)^2 dx$ on the right hand side of $(*)$.

We finally show how *the above a priori estimate implies existence.* The idea is to use a *continuation method:*

Consider

$$\begin{cases} L_t := (1-t)\Delta u + tLu &= f \quad \text{in} \quad \Omega \\ u &= 0 \quad \text{on} \quad \partial\Omega \end{cases}$$

where L is the given elliptic differential operator.

We *assume* that for L_t $(t \in [0,1])$ and boundary values zero we have uniqueness. Set

$$\Sigma := \{t \in [0,1] \mid L_t \text{ is uniquely } \textit{solvable}\} .$$

Since $t = 0 \in \Sigma$, $\sigma \neq \emptyset$.

We shall show that Σ is open and closed and therefore $\Sigma = [0,1]$. Especially we conclude (for $t = 1$) that

$$\begin{cases} Lu &= f \quad \text{in} \quad \Omega \\ u &= 0 \quad \text{on} \quad \partial\Omega \end{cases}$$

has a unique solution.

From the regularity theorem we know that if $A^{\alpha\beta} \in C^{0,\mu}$ and $f \in C^{0,\mu}$ then

$$||u_{xx}||_{C^{0,\mu}(\bar{\Omega})} \leq c\left(||f||_{C^{0,\mu}(\bar{\Omega})} + ||u_{xx}||_{L^2(\Omega)}\right) .$$

Now the first step is to show the

Theorem 3.7 *Suppose for the elliptic operator* $Lu = A_{ij}^{\alpha\beta} u^i_{x_\alpha x_\beta}$ *we have:*

$$\begin{cases} Lu &= 0 \quad \text{in} \quad \Omega \\ u &= 0 \quad \text{on} \quad \partial\Omega \end{cases}$$

has only the zero solution. Then if

$$\begin{cases} Lu &= f \quad \text{in} \quad \Omega \\ u &= 0 \quad \text{on} \quad \partial\Omega \end{cases}$$

we have

$$||u||_{H^{2,2}(\Omega)} \leq c_1 ||f||_{C^{0,\mu}(\bar{\Omega})} .$$

PROOF: Suppose that the theorem is not true, then there exist $A_{ij}^{\alpha\beta(k)}$, $f^{(k)}$ such that for all k

$$A_{ij}^{\alpha\beta(k)}\xi_\alpha\xi_\beta \geq \nu|\xi|^2 \; ; \quad ||A^{(k)}||_{C^{0,\mu}(\bar{\Omega})} \leq M$$

and $||f^{(k)}||_{C^{0,\mu}(\Omega)} \longrightarrow 0 \; (k \to \infty)$.

If $u^{(k)}$ is a solution with $f^{(k)}$ and $||u^{(k)}||_{H^{2,2}(\Omega)} = 1$ then $u^{(k)} \to u \; (k \to \infty)$ and $Lu = 0$ on Ω and $u = 0$ on $\partial\Omega$ i.e. $u = 0$ that is a contradiction.

Remark: Uniqueness of second order equations follows from Hopf's maximum-principle.

Suppose u is a solution of

$$\begin{aligned} A^{\alpha\beta}u_{x_\alpha x_\beta} &= 0 \quad \text{on} \quad \Omega \\ u &= 0 \quad \text{on} \quad \partial\Omega \; . \end{aligned}$$

If u has a maximum point in $x_0 \in \Omega$ then the Hessian $\{u_{x_\alpha x_\beta}\} \leq 0$ and $v := u + \varepsilon|x - x_0|^2$ has still a maximum point in x_0 (for ε sufficiently small). But then $A^{\alpha\beta}v_{x_\alpha x_\beta} = A^{\alpha\beta}u_{x_\alpha x_\beta} + \varepsilon$ const. > 0 and that is a contradiction. One concludes that $u = 0$.

As $0 = t \in \Sigma$ we have

$$||u_{xx}||_{L^2(\Omega)} \leq c||f||_{C^{0,\mu}(\bar{\Omega})}$$

and therefore

$$||u_{xx}||C^{0,\mu}(\bar{\Omega}) \leq ||f||_{C^{0,\mu}(\bar{\Omega})}$$

by the a-priori estimate p. 53.

We show that Σ is *closed:*

Let $t_k \in \Sigma$ and $t_k \to t \; (k \to \infty)$ with

$$\begin{cases} L_{t_k}u^{(k)} &= f \quad \text{on} \quad \Omega \\ u^{(k)} &= 0 \quad \text{on} \quad \partial\Omega \end{cases}$$

because $||u_{xx}^{(k)}||_{C^{0,\mu}(\bar{\Omega})} \leq c||f||_{C^{0,\mu}(\bar{\Omega})}$ we have $u^{(k)} \xrightarrow{C^2(\bar{\Omega})} u$ and

$$\begin{cases} L_t u &= f \quad \text{on} \quad \Omega \\ u &= 0 \quad \text{on} \quad \partial\Omega \end{cases}$$

therefore $t \in \Sigma$.

Now we show that Σ is also *open:*

Let $t_0 \in \Sigma$. Then there exists u such that

$$\begin{cases} L_{t_0}u &= f \quad \text{on} \quad \Omega \\ u &= 0 \quad \text{on} \quad \partial\Omega \; . \end{cases}$$

For $w \in C^{2,\mu}(\bar{\Omega})$ there exists a unique u_w such that

$$\begin{cases} L_{t_0} u_w &= (L_{t_0} - L_t)w + f \quad \text{on} \quad \Omega \\ u_w &= 0 \quad \text{on} \quad \partial\Omega \ . \end{cases}$$

We conclude that

$$||u_w||_{C^{2,\mu}(\bar{\Omega})} \leq c|t - t_0| \; ||w||_{C^{2,\mu}(\bar{\Omega})} + c||f||_{C^{2,\mu}(\bar{\Omega})}$$

and

$$||u_{w_1} - u_{w_2}||_{C^{2,\mu}} \leq c|t - t_0| \; ||w_1 - w_2||_{C^{2,\mu}} \ .$$

If we choose $|t - t_0| =: \delta < 1/c$, we see that the operator

$$\begin{array}{rcl} T : C^{0,\mu}(\bar{\Omega}) & \longrightarrow & C^{0,\mu}(\bar{\Omega}) \\ w \longmapsto & & u_w \end{array}$$

is a contradiction and therefore has a fixpoint which just means that $(t_0-\delta, t_0+\delta) \subset \Sigma$, i.e. that Σ is open.

3.5 A few simple applications to nonlinear equations

Here we want to show how the technique used in the linear case extends to some nonlinear cases. First we show that "lower-order-terms" are irrelevant for the regularity. For example consider

(+) $$\int_\Omega DuD\phi dx = \int_\Omega a(u)D\phi dx \text{ for } \forall\phi \in H_0^1$$

If $\int_{B_R} DvD\phi dx = 0$ for $\forall\phi \in H_0^1$ and $v = u$ on ∂B_R then

$$\begin{aligned} \int_{B_\rho} |Du|^2 dx &\leq c(\rho/R)^n \int_{B_R} |Du|^2 dx + c \int_{B_R} |D(u-v)|^2 dx \\ &\leq c(\rho/R)^n \int_{B_R} |Du|^2 dx + c \int_{B_R} |a(u)|^2 dx \ . \end{aligned}$$

Now if $|a(u)| \leq |u|^\alpha$, then the last integral is estimated by

$$\int_{B_R} |u|^{2\alpha} dx \leq \left(\int_{B_R} |u|^{2*} \right)^{2\alpha/2*} R^{n(1-2\alpha/2*)}$$

so that u is Hölder-continuous provided $\alpha < \frac{2}{n-2}$.

The same result holds if we have instead of (+)

$$\int_\Omega DuD\phi dx = \int_\Omega a(u)D\phi dx + \int_\Omega a(Du)\phi dx$$

with $a(Du) \le |Du|^{1-2/n}$ or $a(Du) \le |Du|^{2-\varepsilon}$ and $|u| \le M = \text{const.}$

The above idea works also if we study small perturbations of a system with constant coefficients:

For

$$\int_\Omega A_{ij}^{\alpha\beta}(x) D_\alpha u^i D_\beta \phi^j dx = 0$$

with

$$A_{ij}^{\alpha\beta} \in L^\infty(\Omega) \quad \text{and} \quad \left| A_{ij}^{\alpha\beta}(v) - \delta_{\alpha\beta}\delta_{ij} \right| \le \varepsilon_0$$

one finds

$$\int_{B_\rho} |Du|^2 dx \le c(\rho/R)^n \int_{B_R} |Du|^2 dx + c \int_{B_R} \underbrace{|\delta_1\delta_2 - A|}_{\le \varepsilon_0} Du^2 dx$$

because

$$\int_\Omega D_\alpha u^i D_\beta \phi^j dx = \int_\Omega (\delta_{\alpha\beta}\delta_{ij} - A_{ij}^{\alpha\beta}) D_\alpha u^i D_\beta \phi^j dx \ .$$

So we can apply the lemma p. 44 and conclude that Du is Hölder-continuous. Finally we mention the

Theorem 3.8 *Let $u \in H^{1,2}(\Omega, \mathbb{R}^N)$ be a minimizer of*

$$\int_\Omega F(Du) dx$$

and let us assume that

$$\frac{F(tp)}{t^2} \to |p|^2 \qquad \text{for } t \to \infty$$

then $u \in C^{0,\mu}(\Omega)$.

Proof: We consider the problem

$$\int_{B_R} |DH|^2 dx \to \text{ minimize}, \qquad H = u \quad \text{on } \partial B_R$$

(H stands for harmonic function!)

then

$$\int_{B_\rho} |Du|^2 dx \le c(\rho/R)^n \int_{B_R} |Du|^2 dx + \int_{B_R} |Du - DH|^2 dx \ .$$

Now

$$\begin{aligned}
\int_{B_R} D(u-H)D(u-H)dx &= \int_{B_R} DuD(u-H)dx - \int_{B_R} DHD(u-H)dx \\
&= \int_{B_R} |Du|^2 dx - \int_{B_R} DuDHdx \\
&= \int_{B_R} |Du|^2 dx - \int_{B_R} (D(u-H)+DH)DHdx \\
&= \int_{B_R} |Du|^2 dx - \int_{B_R} |DH|^2 dx \\
&= \int_{B_R} |Du|^2 dx + \underbrace{\int_{B_R} F(Du)dx - \int_{B_R} F(DH)dx}_{\le 0 (u \text{ is a minimizer})} \\
&\quad - \int_{B_R} F(Du)dx + \int_{B_R} F(DH)dx - \int_{B_R} |DH|^2 dx \ .
\end{aligned}$$

By the growth assumption on F and because

$$|p|^2 - F(p) = |p|^2 \left(1 - \frac{F(p)}{|p|^2}\right)$$

we find

$$\int_{B_R} |D(u-H)|^2 dx \le \varepsilon \int_{B_R} |Du|^2 dx + cR^n \ .$$

So we can apply the lemma p. 44 and conclude that Du is Hölder-continuous.

Chapter 4

L^p-theory

In this chapter we shall prove interior and global L^p-estimates for solutions of linear systems with continuous coefficients.

These estimates are usually derived from the well-known Calderon-Zygmund theorem on singular integrals [6]. Instead of this we follow here Campanato-Stampachia [10], presenting a proof which is free from "potential theory" and relies on an interpolation-theorem due to Stampacchia [69].

In the first section we describe some basic facts on the integrability of functions: weak estimates, Marcinkiewicz-interpolation-theorem, Hardy-Littlewood maximal function and -maximal theorem and Lebesgue-differentiation theorem. Then we present the Calderon-Zygmund-covering argument which plays a central role in the following.

In section 2 we discuss the John-Nirenberg-space BMO [47] and we prove the so-called John-Nirenberg-lemma [47]; moreover we state a sufficient condition in order that a function is L^p-weak (see [47]).

This allows us to prove Stampacchia's interpolation theorem [69], [9].

Finally in section 3 we show how L^p-estimates easily follow from Stampacchia's theorem.

4.1 Some preliminaries

For $f \in L^1(\Omega)$ and $t \geq 0$ we set

$$A_t(f) := \{x \in \Omega \mid |f(x)| > t\} \quad .$$

The *distribution function* λ_f of f is defined as the Lebesgue-measure of $A_t(f)$:

$$\lambda_f(t) := |A_t(f)| := \text{meas}\,\{x \in \Omega \mid |f(x)| > t\} \quad .$$

Note that: $f \in L^\infty(\Omega) \Leftrightarrow ||f||_\infty = \inf\{t \mid \lambda_f(t) = 0\}$ (just observe that $\lambda_f(t) = 0$ if $t \geq ||f||_\infty$ and $\lambda_f(t) > 0$ if $t < ||f||_\infty$).

We have

Proposition 4.1 *(1)* $\int\limits_\Omega |f|^p dx = p \int\limits_0^\infty s^{p-1} |A_s(f)|\, ds$

2) $\int\limits_{A_t(f)} |f|^p dx = p \int\limits_t^\infty s^{p-1} |A_s(f)|\, dx + t^p |A_t(f)|$

PROOF:

1) $\int\limits_\Omega |f|^p dx = \int\limits_\Omega dx \int\limits_0^{|f(x)|} pt^{p-1} dt = \int\limits_\Omega dx \int\limits_0^\infty pt^{p-1} \chi_{A_t} dt = p \int\limits_0^\infty t^{p-1} \int\limits_\Omega \chi_{A_t} dx\, dt$

2) apply 1) to $\tilde{f} = \max(|f|^p - t^p, 0)$.

For $f \in L^p(\Omega)$ we have

$$t^p \operatorname{meas}\{x \in \Omega \mid |f(x)| > t\} \leq \int\limits_{A_t} |f(x)|^p dx \leq \int\limits_\Omega |f(x)|^p dx$$

thus $\lambda_f(t) \leq t^{-p} ||f||^p_{L^p(\Omega)}$. This is called a *weak-L^p-estimate.* We define

$$||f||_{L^p_w(\Omega)} := \inf\left\{A \mid \lambda_f(t) \leq t^{-p} A^p\right\}$$

and $f \in L^p_w(\Omega)$ iff $||f||_{L^p_w(\Omega)} < \infty$. Observe that $||\ \ ||_{L^p_w}$ is *not* a norm!

We have $L^p(\Omega) \subsetneq L^p_w(\Omega) \subset L^q(\Omega)$ for $\forall q < p$, $1 \leq q$ which follows from 1) in the proposition above:

$$\begin{aligned} \int\limits_\Omega |f|^q dx &= q \int\limits_0^\infty t^{q-1} |A_s(f)|\, dt \\ &\leq q \int\limits_0^1 t^{q-1} |A_t| dt + q \int\limits_1^\infty t^{q-1} |A_t| dt \\ &\leq q|\Omega| + A^p q \int\limits_1^\infty t^{q-1-p} dt < \infty\ . \end{aligned}$$

a) **Marcinkiewicz-interpolation theorem.**
Let $T : L^p \to L^q$ be a *quasilinear map* (i.e. $|T(f+g)| \leq Q(|Tf| + |Tg|)$. We say that T is a map of *weak-type* (p,q) if $||Tf||_{L^q_w(\Omega)} \leq A||f||_{L^p(\Omega)}$; and of *strong type* (p,q) if $||Tf||_{L^q(\Omega)} \leq A||f||_{L^p(\Omega)}$ (for $p = q = \infty$ "weak" means strong).

Theorem 4.1 *Suppose that T is a quasilinear map of weak-type (1,1) and of strong type (∞,∞).*
Then, for $1 < p < \infty$, T is strong-type (p,p) (i.e. T maps $L^p(\Omega)$ into $L^p(\Omega)$ continuously).

PROOF: We write $f = g + h$ with $g(x) = \begin{cases} f(x) & \text{if } |f| < \gamma s \\ 0 & \text{otherwise} \end{cases}$ (where γ is to be chosen in the sequel).
Clearly $g \in L^\infty$ and $||g||_\infty \leq \gamma s$.

As T is quasilinear we can estimate

$$\begin{aligned} &\text{meas}\,\{x \in \Omega \mid |Tf(x)| > s\} \\ &\leq \text{meas}\left\{x \in \Omega \mid |Tg(x)| > \tfrac{s}{2Q}\right\} + \text{meas}\left\{x \in \Omega \mid |Th| > \tfrac{s}{2Q}\right\} \end{aligned}$$

Now the first number on the right is zero if γ is sufficiently large:

We have $||Tg||_\infty \leq A_\infty ||g||_\infty \leq A_\infty \gamma s$ and we can choose $\gamma = \frac{1}{2QA_\infty}$.

For the second term we find

$$\text{meas}\left\{x \in \Omega \mid |Th| > \frac{s}{2Q}\right\} \leq \frac{2A_1 Q}{s} \int\limits_{\{|f| \geq \gamma s\} \cap \Omega} |f| dx$$

because $||Th||_{L^1_w} \leq A_1 ||h||_{L^1}$.
We finally compute

$$\begin{aligned} \int\limits_\Omega |Tf|^p dx &= p \int\limits_0^\infty s^{p-1} \text{meas}\,\{x \mid |Tf(x)| > s\}\, dx \\ &\leq 2A_1 Qp \int\limits_0^\infty s^{p-2} \int\limits_{\{|f| \geq \gamma s\} \cap \Omega} |f| dx \\ &= 2A_1 Qp \int\limits_0^\infty s^{p-2} \left\{ \int\limits_{\gamma s}^\infty \text{meas}\,\{x \mid |f(x)| > t\}\, dt + \gamma s \,|A_{\gamma s}(f)| \right\} dx \\ &= 2A_1 Qp \left\{ \int\limits_0^\infty s^{p-2} \int\limits_{\gamma s}^\infty |A_t| dt ds + \int\limits_0^\infty s^{p-2} \gamma s \,|A_{\gamma s}(f)|\, dx \right\} \\ &= 2A_1 Qp \left\{ \int\limits_0^\infty dt |A_t| \int\limits_0^{t/\gamma} s^{p-2} dx + \gamma^{1-p} \int\limits_0^\infty t^{p-1} |A_t| dt \right\} \\ &\leq c \int\limits_\Omega |f|^p dx \ . \end{aligned}$$

Remark: The theorem actually holds in the following form:

If T is of weak-type (p_0, p_0) and also of weak-type (q_0, q_0) for some $p_0 < q_0$, then T is of strong-type (p, p) for all p with $p_0 < p < q_0$.

b) **Hardy-Littlewood maximal function.**
By $Q(x, r)$ we shall design in the following a curve with sides of length $2r$ parallel to the coordinate axes and with center x.

For $f \in L^1_{\text{loc}}(\mathbb{R}^n)$ we set

$$Mf(x) := \sup_{r>0} \frac{1}{|Q(x,r)|} \int_{Q(x,r)} |f(y)|dy$$

Mf is called the *(centered) Hardy-Littlewood maximal function.*

We state some elementary properties:

1) Mf is measurable (in fact $\{x/Mf(x) > t\}$ is open).

2) M is sublinear.

3) If $f \in L^\infty$ then $||Mf||_\infty \leq ||f||_\infty$.

4) $f \in L^1 \not\Rightarrow Mf \in L^1$

We have e.g. that for $f := \chi_{[0,1]}$, Mf behaves as $\frac{1}{x} \notin L^1$; or for $f := x^{-1} \log^{-2} x$, that $Mf \geq x^{-1} \log^{-1} x \notin L^1((0, 1/2))$.

Hardy-Littlewood maximal-theorem. *If $f \in L^1(\mathbb{R}^n)$ then for all $s > 0$*

$$|A_s(Mf)| \leq \frac{c(n)}{s} ||f||_{L^1(\mathbb{R}^n)} , \quad \text{i.e. } ||Mf||_{L^1_w} \leq c(n)||f||_{L^1} .$$

PROOF: For $\forall x \in \{x | Mf(x) > s\}$ there exists $Q(x, r(x))$, such that

$$s < \int_{Q(x,r(x))} |f(y)|dy$$

i.e. $|Q(x, r(x))| \leq \frac{1}{s} \int_{Q(x,r(x))} |f(y)|dy$.

$\{Q(x, r(x))\}$ is a covering of $\{x \mid |Mf(x)| > s\}$. In order to finish the proof it suffices to extract a sequence of cubes $Q(x_i, r(x_i))$, such that

1) $|A_s(Mf)| \leq c(n) \sum_i |Q(x_i, r(x_i))|$.

2) the $Q(x_i, r(x_i))$ are disjoint or overlap at every point only finitely may times.

That such a sequence exists just follows from

Besicovitch's covering theorem. *If $r : E \to \mathbb{R}_+$ is a bounded, positive, measurable function defined on the measurable set E and $\ell > 1$, then there exists an at most countable family of points $x_i \in E$ such that*

(i) $E \subset \bigcup_i Q(x_i), r(x_i)$

(ii) At every point of E the cubes $Q(x_i, r(x_i))$ overlap at most 2^n times.

(iii) The cubes $Q(x_i, (2\ell)^{-1} r(x_i))$ are mutually disjoint.

For a proof we refer to [71].

Consequences of the Hardy-Littlewood maximal-theorem.

1) *If $f \in L^p(\mathbb{R}^n)$ with $1 \leq p \leq \infty$ then $Mf(x) < \infty$ for a.e. x.*
PROOF: If $p = \infty$ this is just the property 3) p. 62. If $p = 1$ we have

$$\{x \in \mathbb{R}^n \mid Mf(x) = \infty\} = \bigcap_{s>0} \{x \in \mathbb{R}^n \mid Mf(x) > s\}$$

and as meas $\{x \in \mathbb{R}^n \mid Mf(x) > s\} < 1/s$, we conclude that $Mf(x) < \infty$ for a.e. x.

For arbitrary $p \in]1, \infty[$ just observe that $f \in L^p$ can be written as $f = f_1 + f_2$ where $f_1 \in L^1$ and $f_2 \in L^\infty$.

2) *By property 3) p. 62 M is of strong type (∞, ∞) and by Hardy-Littlewood maximal-theorem M is of weak-type (1,1), thus we conclude from Marcinkiewicz's theorem (p. 60) that M is of strong-type (p, p) for $1 < p \leq \infty$.*

3) *For a.e. x holds $|f(x)| \leq Mf(x)$ (see the following theorem) hence $||f||_{L^p} \leq ||Mf||_{L^p}$ and $Mf \in L^p$ implies $f \in L^p$.*

c) **Lebesgue-differentiation-theorem.**
Suppose that $f \in L^1(\mathbb{R}^n)$ then for a.e. $x \in \mathbb{R}^n$ we have

$$f(x) = \lim_{\rho \to 0} \fint_{Q(x,\rho)} f(y)dy \quad .$$

PROOF: It is not difficult to show that $\fint f(y)dy$ converges to f in $L^1(\mathbb{R}^n)$ if $\rho \to 0$. So it is sufficient to show that

$$\Omega f(x) := \limsup_{\rho \to 0} \fint_{Q(x,\rho)} f(y)dy - \liminf_{\rho \to 0} \fint_{Q(x,\rho)} f(y)dy = 0 \qquad \text{for a.e. } x \ .$$

First it is clear that if g is continuous, then $\Omega g = 0$. Now continuous functions are dense in $L^1(\mathbb{R}^n)$, i.e. for $\forall \varepsilon > 0\ \exists g \in C_0^\infty(\mathbb{R}^n)$ such that $||f - g||_{L^1} < \varepsilon$.

Suppose that $\Omega f(x) > s$ for $s > 0$, then, as $f = g + (f - g) =: g + h$, this implies $\Omega h > s$. On the other hand $\Omega h \leq 2Mh$. Therefore

$$\{x \in \mathbb{R}^n \mid \Omega f(x) > s\} \subset \{x \in \mathbb{R}^n \mid Mh(x) > s/2\}$$

and by the Hardy-Littlewood-theorem

$$\operatorname{meas}\{x \in \mathbb{R}^n \mid \Omega f(x) > s\} \leq \left|A_{s/2}(Mh)\right| \leq \frac{2c\varepsilon}{s}$$

this goes to zero as ε goes to zero and as $s > 0$ is arbitrary, the theorem is proved.

To get this result we used

1) Fubini's theorem and the absolute continuity of the integral
2) that the theorem holds for a dense subset
3) the Hardy-Littlewood-maximal-theorem

Now we can ask the following *question:*

For $x \in \mathbb{R}^n$ we consider $\mathcal{F}_x$ = family of sets whose elements "shrink" around x. Is it true that

$$f(x) = \lim_{\substack{F \in \mathcal{F}_x \\ \operatorname{diam} F \to 0}} \fint_F f(y)dy \quad ?$$

Clearly a positive answer would be a consequence of some maximal theorem like: For $f \in L^1(\mathbb{R}^1)$ and $\forall s$

$$\operatorname{meas}\{x \in \mathbb{R}^n \mid M_{\mathcal{F}} f(x) > s\} \leq \frac{c}{s}||f||_{L^1(\mathbb{R}^n)}$$

where $M_{\mathcal{F}} f(x) := \sup\limits_{F \in \mathcal{F}_x} \fint_F |f(y)|dy$.

To achieve such a result we mainly have two possibilities:

(a) We may try to extend Besicovic's covering theorem to our family $\mathcal{F}_x$.

(b) We may try to recover the weak estimate form the maximality theorem.

(b) can be accomplished in the following situation:

We call a family $\mathcal{F}$ *regular* if for $\forall F \in \mathcal{F}$ there exists a cube Q centered at $x = 0$ such that $F \subset Q$, $c|Q| < |F|$ where c is an absolute constant.

For $x \in \mathbb{R}^n$ set $\mathcal{F}_x := x + \mathcal{F}$ then clearly

$$M_{\mathcal{F}} f(x) \leq \frac{1}{c} M f(x) \quad .$$

So we can conclude:

If $\mathcal{F}$ is a regular family the answer to our question is yes — in particular we have the

Theorem 4.2 *If $f \in L^1(\mathbb{R}^n)$ then for a.e. $x \in \mathbb{R}^n$*

$$f(x) = \lim_{\substack{|Q|\to 0 \\ x\in Q}} \fint_Q f(y)dy \ .$$

A simple consequence is also the

Theorem 4.3 (Riesz) *$f \in L^p(Q)$ if and only if*

$$\sup\left\{\sum_i |Q_i| \left(\fint_{Q_i} |f|\right)^p\right\} < \infty$$

where the supremum is to be taken over all decompositions of Q with $Q = \bigcup_i Q_i$.

d) **Calderon-Zygmund-covering argument.**
For $f \in L^1(Q)$, $f \geq 0$ and $\alpha \in \mathbb{R}_+$ we assume that $\fint_Q f(x)dx < \alpha$.

Then there exists a family of cubes $\{Q_j\}$, such that

(i) the Q_j have disjoint interiors and sides parallel to the sides of Q

(ii) $\alpha < \fint_{Q_j} f(x)dx \leq 2^n\alpha$

(iii) $f(x) \leq \alpha$ for a.e. $x \in Q \setminus \bigcup_j Q_j$

(iv) $\sum_j |Q_j| \leq \frac{1}{\alpha}\int_Q f(x)dx$. ($|Q_j|$= volume of Q_j)

PROOF: We divide Q into 2^n equal cubes Q_i. If $\fint_{Q_i} f(x)dx > \alpha$ we take Q_i to the family we are looking for.

If $\fint_{Q_i} f(x)dx \leq \alpha$ we continue to divide. This construction implies (i) and (iii). For $x \in Q\setminus\bigcup_j Q_j$ (ii) immediately follows from Lebesgue's differentiation-theorem.

4.2 The spaces of functions of bounded mean-oscillation = BMO-spaces = John-Nirenberg-spaces

Let Q_0 be a cube in $\mathbb{R}^n$ and u a measurable function on Q_0. We define:

$$u \in BMO(Q_0)$$

if and only if

$$|u|_{*,Q_0} := \sup_{Q} \fint_{Q\cap Q_0} |u - u_{Q\cap Q_0}| dx \ .$$

It is easy to see that we can also take as definition

$$|u|_{*,Q_0} = \sup_{Q\subset Q_0} \fint_{Q} |u - u_Q| dx$$

and that $u \in BMO(Q_0)$, if and only if $u \in \mathcal{L}^{1,n}(Q_0)$.

We have the

Theorem 4.4 (John/Nirenberg) *There exist constants $c_1 > 0$ and $c_2 > 0$ depending only on n, such that the following holds:*

If $u \in BMO(Q_0)$, then for all $Q \subset Q_0$

$$\text{meas}\,\{x \in Q_0 \mid (u - u_Q)(x) > t\} \le c_1 \exp\left(\frac{-c_2}{|u|_{*,Q_0}} t\right) |Q| \ .$$

PROOF: It is sufficient to give the proof for $Q = Q_0$ (because $u \in BMO(Q_0) \Rightarrow u \in BMO(Q)$ for $Q \subset Q_0$). Moreover the inequality is homogeneous and so we can assume that $|u|_{*,Q_0} = 1$.

For $\alpha > 1 = \sup\limits_{Q\subset Q_0} \fint_Q |u - u_Q| dx$ we use Calderon-Zygmund-argument to get a sequence $\{Q_j^{(1)}\}$ with

(i) $\alpha < \fint_{Q_j^{(1)}} |u - u_{Q_0}| dx \le 2^n \alpha$

(ii) $|u - u_{Q_0}| \le \alpha$ on $Q_0 \setminus \bigcup_j Q_j^{(1)}$.

Then we also have

$$\left|u_{Q_j^{(1)}} - u_{Q_0}\right| \le \fint_{Q_j^{(1)}} |u - u_{Q_0}| \le 2^n \alpha$$

and $\sum\limits_j |Q_j^{(1)}| \le \frac{1}{\alpha} \int\limits_{Q_0} |u - u_{Q_0}| dx.$

Take *one* of the $Q_j^{(1)}$. Then as $|u|_{*,Q_0} = 1$ we have $\fint_{Q_j^{(1)}} \left|u - u_{Q_j^{(1)}}\right| \le 1 < \alpha$ and we can apply Calderon-Zygmund argument again. Thus there exists a sequence $\{Q_j^{(2)}\}$ with

$$|u(x) - u_{Q_0}| \le \left|u(x) - u_{Q_j^{(1)}}\right| + |u_{Q_j^{(1)}} - u_{Q_0}| \le 2 \cdot 2^n \alpha$$

for a.e. x on $Q_0 \setminus \cup Q_j^{(2)}$ and

$$\begin{aligned}
\sum_j |Q_j^{(2)}| &\le \frac{1}{\alpha} \sum_j \int_{Q_j^{(1)}} \left|u - u_{Q_j^{(1)}}\right| \\
&\le \frac{1}{\alpha} \sum_j |Q_j^{(1)}| \qquad (u \in BMO \text{ and } |u|_* = 1) \\
&\le \frac{1}{\alpha^2} \int_{Q_0} |u - u_{Q_0}| dx \\
&\le \frac{1}{\alpha^2} |Q_0| \ .
\end{aligned}$$

Repeating this argument one gets by induction: For all $k \ge 1$ there exists a sequence $\{Q_j^{(k)}\}$ such that

$$|u - u_{Q_0}| \le k 2^n \alpha \quad \text{a.e. on } Q_0 \setminus \cup_j Q_j^{(k)}$$

and

$$\sum_j |Q_j^{(k)}| \le \frac{1}{\alpha^k} \int_{Q_0} |u - u_{Q_0}| dx \ .$$

Now for $t \in \mathbb{R}_+$ either $0 < t < 2^n \alpha$ or there exists $k \ge 1$, such that $2^n \alpha k < t < 2^n \alpha (k+1)$.

In the first case

$$\text{meas}\,\{x \in Q_0 \mid |u(x) - u_{Q_0}| > t\} \le |Q_0| \le |Q_0| e^{-At} e^{2^n A\alpha}$$

$(o \le 2^n \alpha - t \Rightarrow 1 \le e^{A(2^n \alpha - t)}$ for some $A > 0)$;

in the second case

$$\begin{aligned}
&\text{meas}\,\{x \in Q_0 \mid |u(x) - u_{Q_0}| > t\} \\
&\le \ \text{meas}\,\{x \in Q_0 \mid |u(x) - u_{Q_0}| > 2^n \alpha k\} \\
&\le \ \sum_j |Q_j^{(k)}| \le \frac{1}{\alpha^k} \int_{Q_0} |u - u_{Q_0}| dx \\
&\le \ e^{\left(1 - \frac{t}{2^n \alpha}\right) \log \alpha} \int_{Q_0} |u - u_{Q_0}| dx \le \alpha e^{\frac{-\log \alpha}{2^n \alpha} t} |Q_0| \\
&=: \ \alpha e^{-At} |Q_0| \qquad \left(\text{we used } -k \le 1 - \tfrac{t}{2^n \alpha} \text{ and } \alpha^{-k} = e^{-k \log \alpha}\right).
\end{aligned}$$

Corollary 4.1 *If $u \in BMO(Q_0)$, then $u \in L^p(Q_0)$ for $\forall p > 1$. Moreover*

$$\sup_{Q \subset Q_0} \left(\int_Q |u - u_Q|^p dx \right)^{1/p} \leq c(n,p)|u|_{*,Q_0}.$$

PROOF:

$$\begin{aligned}
\int_Q |u - u_Q|^p dx &= p \int_0^\infty t^{p-1} \operatorname{meas}\{x \in Q \mid |u - u_Q| > t\}\, dt \\
&\leq pc_1 \int_0^\infty t^{p-1} \exp\left(-\frac{c_2}{|u|_{*,Q}} t \right) |Q| dt \\
&\text{change variables} \\
&= pc_1 \left(\frac{c_2}{|u|_{*,Q}} \right)^{-p} |Q| \int_0^\infty e^{-t} t^{p-1} dx \leq c(n,p)|u|^p_{*,Q}
\end{aligned}$$

Corollary 4.2

$$\mathcal{L}^{p,n}(Q_0) \cong BMO(Q_0) \qquad \text{for } \forall p\ .$$

Finally it is not difficult to show that we have the following

Theorem 4.5 (Characterization of BMO)
The following statements are equivalent:

(1) $u \in BMO(Q_0)$

(2) there exist constants c_1, c_2 such that for all $Q \subset Q_0$ and all $s > 0$

$$\operatorname{meas}\{x \in Q \mid |u(x) - u_Q| > s\} \leq c_1 \exp(-c_2 s)|Q|$$

(3) there exist constants c_3, c_4 such that for all $Q \subset Q_0$

$$\int_Q \left(\exp(c_4|u - u_Q|) - 1\right) dx \leq c_3|Q|$$

(4) if $v = \exp(c_4 u)$ then

$$\sup_{Q \subset Q_0} \left(\fint_Q v dx \right) \left(\fint_Q \frac{1}{v} dx \right) \leq c_5\ .$$

(We can take $c_1 = c_3 = c_5 = c(n)$ and $c_2 = 2c_4 = c(n)|u|^{-1}_{*,Q_0}$.)

We shall now prove the following interpolation theorem due to Stampacchia [69].

Theorem 4.6 *Suppose*

$$T : L^p(Q) \to L^p(Q_0)$$

$(1 \le p < \infty)$ *and*

$$T : L^\infty(Q) \to BMO(Q_0)$$

is a bounded linear operator. Then $T : L^q(Q) \to L^q(Q_0)$ *is also continuous for all* q *with* $p < q < \infty$.

For the proof of this theorem we shall use a theorem due to John/Nirenberg [47] which we state without proof:

Theorem 4.7 *Let* Δ *denote the set of all subdivisions* I *of* Q_0 *into cubes* Q_i. *If* $K_p^p(u) := \sup_{I\in\Delta} \sum_{i\in I} |Q_i| \left(\fint_{Q_i} |u - u_{Q_i}| dx \right)^p < \infty$ *then* $u - u_{Q_0} \in L_w^p(Q_0)$, *i.e.*

$$\text{meas}\,\{x \in Q_0 \mid |u(x) - u_{Q_0}| > t\} \le c(n,p) \left(\frac{K_p(u)}{t} \right)^p .$$

Remark: It is not difficult to see that:

$$u \in BMO(Q_0) \Leftrightarrow \lim_{p\to\infty} K_p(u) < \infty$$

and in this case $|u|_{*,Q_0} = \lim_{p\to\infty} K_p(u)$. We now come to the

PROOF OF STAMPACCHIA'S THEOREM (cf. Campanato [9]): We start with a partition $\{Q_k\}$ of Q_0. For $u \in L^p(\Omega)$ we define

$$\tilde{T}(x) := \fint_{Q_k} |Tu - (Tu)_{Q_k}|\, dy \qquad \text{for } x \in Q_k$$

then

1) $\tilde{T}$ is a strong-type (p,p) map:

$$\begin{aligned} ||\tilde{T}u||_{L^p(Q_0)} &= \sum_k |Q_k|^p \left(\fint_{Q_k} |Tu - (Tu)_{Q_k}|\, dx \right)^p \\ &\le \sum_k \int_{Q:k} |Tu - (Tu)_{Q_k}|^p\, dx \\ &\le 2^{p-1} \left\{ 2 \int_{Q_0} |Tu|^p dx \right\} \\ &\le 2^p M_p^p \int_Q |u|^p dx \end{aligned}$$

2) $\tilde{\mathcal{T}}$ is of strong-type (∞, ∞) since $Tu \in BMO(Q_0)$

3) $\tilde{\mathcal{T}}$ is quasilinear.

From 1), 2) and 3) we conclude by Marcinkiewicz's theorem that $\tilde{\mathcal{T}}$ is a bounded map from $L^r(Q)$ into $L^r(Q_0)$ for $\forall r \in (p, \infty)$:

$$||\tilde{\mathcal{T}}||_{L^r(Q_0)} \leq c(p, r, M_\infty, M_p)||u||_{L^r(Q)} \ .$$

Now

$$K_r(\tilde{\mathcal{T}}u) = \left\{ \sup_{I\in\Delta} \sum_{i\in I} |Q_i| \left(\fint_{Q_i} |Tu - (Tu)_{Q_i}|\, dx \right)^r \right\}^{1/r} \leq \text{constant } ||u||_{L^r(Q_0)} \ ,$$

hence by the above mentioned theorem $\tilde{\mathcal{T}}$ is of weak type (r, r) and again by Marcinkiewicz (but this time applied to T) the theorem is proved.

Remark: A maybe clearer proof of the interpolation-theorem can be deduced from a result of Fefferman and Stein:

For $u \in L^1_{\text{loc}}(\mathbb{R}^n)$ we define the *sharp function* of u by

$$u^\sharp(x) := \sup_{x\in Q} \int_Q |u - u_Q| dy \ ;$$

then we have the

Theorem 4.8 (Fefferman/Stein) *For* $p > 1$ *holds*

$$u \in L^p(\mathbb{R}^n) \Leftrightarrow u^\sharp \in L^p(\mathbb{R}^n)$$

and $||u||_{L^p(\mathbb{R}^n)} \leq c(n, p)||u^\sharp||_{L^p(\mathbb{R}^n)}$.

This result can be localized: If we set

$$u^\sharp_{Q_0}(x) := \sup_{\substack{Q\subset Q_0 \\ x\in Q}} \fint_Q |u - u_Q| dy$$

then for $p > 1$: $u \in L^p(Q_0) \Leftrightarrow u^\sharp_{Q_0} \in L^p(Q_o)$. Under the assumptions of the interpolation theorem of Stampacchia the map $u \mapsto (Tu)^\sharp$ is of type (p, p) and (∞, ∞), then it is also of type (q, q) for all q.

4.3 L^p-theory

We first look at a solution of the following elliptic system with *constant coefficients:*

$$\begin{cases} \int A_{ij}^{\alpha\beta} D_\alpha u^i D_\beta \phi^j dx &= \int\limits_\Omega f_\alpha^i D_\alpha \phi^i dx \\ u &= 0 \qquad \text{on } \partial\Omega \,. \end{cases}$$

We set $T(f) := Du$, then T is a continuous linear operator

$$T : L^2(\Omega) \to L^2(\Omega)$$

respectively

$$T : L^\infty(\Omega) \to BMO(\Omega)$$

(cf. chapter 3!).

From the interpolation theorem of section 4.2 then follows:

If $f \in L^p(\Omega)$ then $Du \in L^p(\Omega)$ for $2 < p < \infty$ and $||Du||_{L^p(\Omega)} \le c||f||_{L^p(\Omega)}$.

For $\Omega = B_R$ we have $||Du||_{L^p(B_R)} \le c||f||_{L^p(B_R)}$, where the constant c is *independent of R.*

We now generalize to the case of *nonconstant coefficients:*

Look at a local solution of

$$\int\limits_{B_R} A_{ij}^{\alpha\beta}(x) D_\alpha u^i D_\beta \phi^j dx = \int\limits_{B_R} (g^i \phi^i + f_\alpha^i D_\alpha \phi^i) dx \qquad \text{for all } \phi \in H_0^1(B_R).$$

We write

$$\int\limits_{B_R(x_0)} A(x_0) Du D\phi = \int\limits_{B_R(x_0)} \{[A(x_0) - A(x)]\, Du D\phi + g\phi + fD\phi\} \;\; ;$$

as u is not zero on the boundary we multiply it by $\eta \in C_0^\infty(B_R)$ and we find:

$$\begin{aligned} &\int\limits_{B_R} A(x_0) D(u\eta) D\phi \\ &= \int\limits_{B_R} [A(x_0) - A(x)]\, D(u\eta) D\phi + \int\limits_{B_R} A(x) D(u\eta) D\phi \\ &= \int\limits_{B_R} [A(x_0) - A(x)]\, D(u\eta) D\phi \\ &\quad + \int\limits_{B_R} A(x) Du\eta D\phi + \int\limits_{B_R} A(x) u D\eta D\phi \end{aligned}$$

$$\begin{aligned}
&= \int_{B_R} [A(x_0) - A(x)]\, D(u\eta) D\phi \qquad (*)\\
&\quad + \int_{B_R} A(x) Du D(\eta\phi) - \int_{B_R} A(x) Du D\eta \phi + \int_{B_R} A(x) u D\eta D\phi \\
&= \int_{B_R} [A(x_0) - A(x)]\, D(u\eta) D\phi + \int_{B_R} g\eta\phi \\
&\quad + \int_{B_R} f D(\eta\phi) - \int_{B_R} A(x) Du D\eta\phi + \int_{B_R} A(x) u D\eta D\phi \\
&= \int_{B_R} [A(x_0) - A(x)]\, D(u\eta) D\phi \\
&\quad + \int_{B_R} \phi \{g\eta + f D\eta - A(x) Du D\eta\} + \int_{B_R} D\phi\{f\eta + A(x) u D\eta\} \\
&=: \int_{B_R} [A(x_0) - A(x)]\, D(u\eta) D\phi \\
&\quad + \int_{B_R} G\phi + \int_{B_R} F D\phi \ .
\end{aligned}$$

Let $w \in H_0^{1,2}(B_R, \mathbb{R}^N)$ be the solution of

$$\int_{B_R} Dw D\phi = \int_{B_R} [g + [A(x_0) - A(x)]\, Du D\eta]\, \phi \qquad \text{for } \forall \phi \in H_0^{1,2}(B_r)$$

(note that u is *fixed!*)

We differentiate and conclude from the remarks at the beginning:

If $g\eta + f D\eta - A(x) Du D\eta =: G \in L^r(B_R)$ then $D^2 w \in L^r(B_R)$ and $||D^2 w||_{L^r(B_R)} \leq c||G||_{L^r(B_R)}$.

From Sobolev's inequality then follows

$$||Dw||_{L^{r*}(B_R)} \leq c||G||_{L^r(B_R)} \ .$$

Our equation $(*)$ p. 72 becomes

$$(**) \quad \int_{B_R(x_0)} A(x_0) D(u\eta) D\phi = \int_{B_R(x_0)} [A(x_0) - A(x)]\, D(u\eta) D\phi + \int_{B_R(x_0)} \tilde{F} D\phi$$

where $\tilde{F} = Dw + F$.

Assume that $f \in L^p$, $g \in L^q$ (with q such that $q^* = p$) and $Du \in L^m$, then $G \in L^{\min(m,q)}$, $F \in L^{\min(m*,p)}$ by the definition of G and F.

By the above argument we conclude that $Dw \in L^{\min(m,q)^*}$ and therefore $\tilde{F} \in L^{\min(m^*,p)}$.

Now we fix $V \in H^{1,s}(B_R, \mathbb{R}^N)$ and take a solution $v \in H_0^{1,s}$ of the equation

$$\int_{B_R} A(x_0) Dv D\phi = \int_{B_R} [A(x_0) - A(x)]\, DV D\phi + \int_{B_R} \tilde{F} D\phi$$

then

$$||Dv||_{L^{\min(x,m^*,p)}} \leq c\,||[A(x_0) - A(x)]\, DV||_{L^{\min(x,m^*,p)}} + c||\tilde{F}||_{L^{\min(x,m^*,p)}}$$

(see the remarks at the beginning of this section).

We take $s = \min(m^*, p)$. The continuous operator

$$\begin{array}{rcl} T : H_0^{1,s} & \longrightarrow & H_0^{1,s} \\ V & \longmapsto & v \end{array}$$

is a contraction for R sufficiently small:

$$\begin{aligned} ||DT(V_1 - V_2)||_{L^s(B_R)} &\leq c\,||[A(x_0) - A(x)]\, D(V_1 - V_2)||_{L^s(B_R)} \\ &\leq c \sup_{B_R} [A(x_0) - A(x)]\, ||DV_1 - DV_2||_{L^s(B_R)} \,. \end{aligned}$$

Thus it follows that there exists a *unique fixpoint* which is a solution of (**) and of course equals $u\eta$, moreover

$$||D(u\eta)||_{L^s(B_R)} \leq c||\tilde{F}||_{L^s(B_R)} \,.$$

We have proved:

If $f \in L^p(\Omega)$ and $f \in L^{\frac{pn}{n-p}}(\Omega)$ then $Du \in L^p_{\text{loc}}(\Omega)$.

A similar argument permits to get an *L^p-theory in the nonvariational case: Suppose u is a solution of*

$$a^{\alpha\beta}(x) u_{x_\alpha x_\beta} = f \in L^p(\Omega)$$

with $a^{\alpha\beta} \in C^0(\Omega)$, $a^{\alpha\beta}\xi_\alpha\xi_\beta \geq \nu|\xi|^2$ for $\forall \xi \in \mathbb{R}^n$, $\nu > 0$, then $u_{x_\alpha x_\beta} \in L^p_{\text{loc}}(\Omega)$.

Finally it should be clear how to prove by the arguments above the following two results:

(1) *Let u be a weak solution of*

$$\int_\Omega A^{\alpha\beta}(x) D_\alpha u^i D_\beta^j = \int_\Omega g^i \phi^i + f_\alpha^i D_\alpha \phi^i \,, \quad u - \phi \in H_0^{1,p}(\Omega)$$

with $f^i_\alpha \in L^p(\Omega)$, $g \in L^{\frac{np}{n-p}}(\Omega)$, $A^{\alpha\beta} \in C^0(\Omega)$

$$A^{\alpha\beta}\xi_\alpha\xi_\beta\eta_i\eta_j \geq \nu|\xi|^2|\eta|^2 \qquad \text{for } \forall\xi \in \mathbb{R}^n,\ \ \eta \in \mathbb{R}^N,\ \ \nu > 0\ ,$$

then $u \in H^{1,p}(\Omega)$.

(2) *Let* u *be a solution of*

$$a^{\alpha\beta}(x)u_{x_\alpha x_\beta} = f \in L^p(\Omega)\ , \qquad u - \phi \in H^{2,p}_0(\Omega)$$

$$a^{\alpha\beta} \in C^0(\Omega)\ ,\ \ a^{\alpha\beta}\xi_\alpha\xi_\beta \geq \nu|\xi|^2 \qquad \text{for } \forall\xi \in \mathbb{R}^n,\ \ \nu > 0\ ,$$

then $u \in H^{2,p}(\Omega)$.

Chapter 5

Regularity in the scalar case

In this chapter we describe some beautiful results and techniques in the regularity theory for *scalar* minimizers and solutions of second order elliptic equations.

For the sake of simplicity we shall confine ourselves to the study of interior regularity and we shall always assume quadratic growth instead of a general polynomical growth.

In section 1, following De Giorgi [11] (see also [49], [70]), we introduce the De Giorgi class $DG(\Omega)$ and prove that *functions in $DG(\Omega)$ are Hölder-continuous.* This has as a consequence the celebrated De Giorgi-theorem: *Solutions of second order elliptic equations with bounded measurable coefficients are Hölder-continuous.* From this theorem then follows the regularity of minimizers of regular functionals whose integrands do *not* depend on the variable u.

In section 2 we describe *Moser's iteration technique* [57], [58], from which follows a different proof of De Giorgi's theorem and moreover the elegant *Harnack inequality* for solutions of second order elliptic equations with bounded measurable coefficients. Moser's technique relies strongly on the fact that u is a solution of an elliptic equation and it is not clear whether one can extend it in order to prove Harnack's inequality for the functions in the De Giorgi class.

In section 3 we present a slightly different proof of a recent result by Di Benedetto and Trudinger [13]: *Harnack's inequality holds for functions in the De Giorgi class $DG(\Omega)$.* This follows from results due to De Giorgi [11] and a covering lemma by Krylov-Safanov [48].

In section 4 we discuss the notion of *quasi-minima,* introduced by Giaquinta-Giusti [26], [30]. Among other things, see [30], this notion includes minimizers of *non*-convex functionals and solutions of elliptic equations and systems.

We show that *scalar* quasi-minima are Hölder-continuous and that Harnack's inequality holds for them.

Finally in section 5 we prove that, under certain assumptions, minimizers of multiple integrals in the calculus of variations have Hölder-continuous first derivatives. This result is due to Giaquinta-Giusti [31].

5.1 De Giorgi's class and De Giorgi's theorem

Consider the following situation:

For $\mathbb{F}(u) := \int_\Omega F(Du)dx$ we assume that

(i) $F \in C^2(\mathbb{R}^n)$

(ii) $F_{p_\alpha p_\beta}(p)\xi_\alpha\xi_\beta \geq |\xi|^2$ for $\forall \xi \in \mathbb{R}^n$

(iii) $|F_{pp}| \leq L(=\text{constant})$.

If u is a minimizer for $\mathbb{F}$ in $H^{1,2}(\Omega)$ then u is a solution of

$$\int_\Omega F_{p_\alpha}(Du)D_\alpha\phi dx = 0 \qquad \text{for } \forall \phi \in H_0^{1,2}(\Omega) \ ,$$

which implies that $u \in H^{2,2}_{\text{loc}}(\Omega)$ (see 2.4, p. 30). We now look at

$$\int_\Omega F_{p_\alpha p_\beta}(Du)D_\beta(D_s u)D_\alpha\phi dx = 0 \qquad \text{for } \forall \phi \in H_0^{1,2}(\Omega)$$

and consider $F_{p_\alpha p_\beta}(Du(x)) =: A^{\alpha\beta}(x)$ as a function in $L^\infty(\Omega)$. This leads to the following *regularity problem:*

Suppose that $A^{\alpha\beta} \in L^\infty(\Omega)$ and that

$$A^{\alpha\beta}\xi_\alpha\xi_\beta \geq |\xi|^2 \qquad \text{for } \forall \xi \in \mathbb{R}^n \ .$$

If $u \in H^{1,2}(\Omega)$ is a solution of

$$(+) \qquad \int_\Omega A^{\alpha\beta}(x)D_\alpha u D_\beta\phi dx = 0 \qquad \text{for } \forall \phi \in H_0^{1,2}(\Omega) \ ,$$

can we conclude that u is Hölder-continuous? It is our aim to show that this fact is true. This is just the content of De Giorgi's theorem [11] (see also [59].

The *key idea* is to use *Cacciopoli inequalities on level sets of* u, namely we consider

$$\phi := \max(u-k,0)\eta^2 =: (u-k)^+\eta^2$$

for all $k \in \mathbb{R}$ as test functions. Then for the above situation we immediately have (cf. p. 24)

$$\int_{A(k,\rho)} |Du|^2 dx \leq \frac{c}{(R-\rho)^2} \int_{A(k,R)} |u-k|^2 dx \qquad \text{for } \forall k \in \mathbb{R} \ ,$$

where $A(k,r) := \{x \in B_r \mid u(x) > k\}$. This is equivalent to

$$(*) \qquad \int_{B_\rho} \left|D(u-k)^+\right|^2 dx \leq \frac{c}{(R-\rho)^2} \int_{B_R} \left|(u-k)^+\right|^2 dx \qquad \text{for } \forall k \in \mathbb{R} \ .$$

We define the *De Giorgi-class* ($= DG(\Omega)$) *to consist of all* $u \in H^{1,2}(\Omega)$ *which satisfy* $(*)$.

Remarks:

1) If u is a subsolution
(i.e. $\int_\Omega A^{\alpha\beta} D_\alpha u D_\beta \phi dx \leq 0$ for $\forall \phi \in H_0^{1,2}(\Omega), \phi \geq 0$)
then $u \in DG(\Omega)$.

2) If u is a supersolution
(i.e. $\int_\Omega A^{\alpha\beta} D_\alpha u D_\beta \phi dx \geq 0$ for $\forall \phi \in H_0^{1,2}(\Omega)$, $\phi \geq 0$)
then $-u \in DG(\Omega)$.

3) If $u \in DG(\Omega)$ then $u+$ constant $\in DG(\Omega)$.

4) In $(*)$ we have the exponent $p = 2$ but actually everything works as well for any $p > 1$.

Now if $u \in DG(\Omega)$, then clearly

$$\int_{B_{\frac{R+\rho}{2}}} \left|D(u-k)^+\right|^2 dx \leq \frac{c}{(R-\rho)^2} \int_{B_R} \left|D(u-k)^+\right|^2 dx \ ;$$

and if we choose $\eta \in C_0^\infty\left(B_{\frac{R+\rho}{2}}\right)$ with $\eta \equiv 1$ on B_ρ, then

$$\int_{B_{\frac{R+\rho}{2}}} \left|D\left(\eta(u-k)^+\right)\right|^2 dx \leq \frac{c}{(R-\rho)^2} \int_{B_R} \left|D(u-k)^+\right|^2 dx \tag{5.1}$$

this implies (Sobolev-inequality and $\frac{R+\rho}{2} > \rho$!)

$$\left(\int_{B_\rho} \left|(u-k)^+\right|^{2^*} dx\right)^{2/2^*} \leq \frac{c}{(R-\rho)^2} \int_{B_R} \left|(u-k)^+\right|^2 dx \ ; \tag{5.2}$$

moreover by Hölder's inequality

$$\int_{B_\rho} |(u-k)^+|^2\, dx \le \left(\int_{B_\rho} |(u-k)^+|^{2^*}\, dx \right)^{2/2^*} \operatorname{meas}\{x \in B_\rho \mid u(x) > k\}^{1-2/2^*} . \tag{5.3}$$

We conclude from (5.1) that

$$\int_{A(k,\rho)} |Du|^2 dx \le \frac{c}{(R-\rho)^2} \int_{A(k,R)} |u-k|^2 dx \qquad \text{for } \forall k \in \mathbb{R}$$

and by inserting (5.2) into (5.3) that

$$\int_{A(k,\rho)} |u-k|^2 dx \le \frac{c}{(R-\rho)^2} \int_{A(k,R)} |u-k|^2 dx |A(k,R)|^{2/n} \; \left(2^* = \tfrac{2n}{n-2}\; ;\; 1 - \tfrac{2}{2^*} = \tfrac{2}{n}\right) .$$

For $h > k$

$$|h-k|^2 |A(h,\rho)| = \int_{A(h,\rho)} |h-k|^2 dx \le \int_{A(k,\rho)} |u-k|^2 dx .$$

We set

$$\begin{aligned} a(h,\rho) &:= |A(h,\rho)| \\ u(h,\rho) &:= \int_{A(h,\rho)} |u-h|^2 dx \end{aligned}$$

and then write the above inequalities in the form

$$\begin{aligned} u(h,\rho) &\le \frac{c}{(R-\rho)^2} u(k,R) a(k,R)^{2/n} \\ a(h,\rho) &\le \frac{1}{(h-k)^2} u(k,R) . \end{aligned}$$

For some positive numbers ξ and η we find:

$$u(h,\rho)^\xi a(h,\rho)^\eta \le \frac{c^\xi}{(R-\rho)^{2\xi}} \frac{1}{(h-k)^{2\eta}} u(k,R)^{\xi,R} a(k,R)^{2\xi/\eta} .$$

Now we choose ξ and η in such a way that for some θ (we are looking for)

$$\xi + \eta = \theta\xi \qquad \text{and} \quad \frac{2\xi}{\eta} = \theta\eta \; ;$$

then θ must be a positive solution of

$$\theta^2 - \theta - \frac{2}{n} = 0 \qquad \text{i.e. } \theta = 1/2 + \sqrt{1/4 + 2/n} > 1$$

and we can fix $\eta = 1$, $\xi = \frac{n}{2}\theta$.

Setting $\phi(h,\rho) := u(h,\rho)^\xi a(h,\rho)^\eta$ we conclude that for all ρ and R satisfying $\rho < R <$ some R_0, and for all $h > k >$ some $\tilde{k}$ (in our case $\tilde{k} = -\infty$) we have

$$\phi(h,\rho) \leq \frac{c^\xi}{(R-\rho)^{2\xi}(h-k)^{2\eta}} \phi(k,R)^\theta \ , \qquad \theta > 1 \ .$$

Proposition 5.1 *For any $k_0 \geq \tilde{k}$ and $\sigma \in (0,1)$ we have*

$$\phi(k_0 + d, R_0 - \sigma R_0) = 0 \ ,$$

where

$$d^{2\eta} = \frac{2^{(2\xi+2\eta)\frac{\theta}{\theta-1}} c^\xi \phi(k_0, R_0)^{\theta-1}}{\sigma^{2\xi} R_0^{2\xi}} \ .$$

PROOF (hints): Set $k_n := k_0 + d - \frac{d}{2^n}$ and $\rho_n := R_0 - \sigma R_0 + \frac{\sigma k_0}{2^n}$; then by induction

$$\phi(k_n, \rho_n) \leq \frac{\phi(k_0, R_0)}{2^{\mu n}} \ ; \qquad \mu = \frac{2\xi + 2\eta}{\theta - 1} \ .$$

The result follows by taking the limit $n \to \infty$.

As a consequence of this we have for $\sigma = 1/2$

Theorem 5.1 *If $u \in DG(\Omega)$ then*

$$\sup_{s \in B_{R/2}} |u(x)| \leq k_0 + c \left(\frac{1}{R^n} \int_{A(k_0,R)} |u - k_0|^2 dx \right)^{1/2} \left(\frac{A(k_0,R)}{R^n} \right)^{\frac{\theta-1}{2}} \ .$$

PROOF: $\phi(k_0 + d, R_0/2) = 0 = u(k_0 + d, R_0/2)^\xi a(k_0 + d, R_0/2)^n$, thus either $u(k_0 + d, R_0/2) = 0$ or $a(k_0 + d, R_0/2) = 0$ and hence the result.

In particular for $k_0 = 0$ we have the

Theorem 5.2 *If $u \in DG(\Omega)$, then* $\sup_{B_{R/2}} u^+ \leq c \left(\fint_{B_R} |u^+|^2 dx \right)^{1/2}$

and the

Theorem 5.3 *If $-u \in DG(\Omega)$, then* $\sup_{B_{R/2}} u^- \leq c \left(\fint_{B_R} |u^-|^2 dx \right)^{1/2}$ *for some constant c ($u^+ := \max(u,0)$; $u^- := \max(-u,0)$).*

From these two estimates we infer

Theorem 5.4 *If u is a solution of $(+)$ p. 76, then u is bounded and*

$$\sup_{B_{R/2}} |u| \le c \left(\fint_{B_R} |u|^2 dx \right)^{1/2} .$$

Remarks:

1) *For all $\tau, 0 < \tau < 1$, we have*

$$\sup_{B_\tau R} u \le \frac{c}{(1-\tau)^{n/2}} \left(\fint_{B_R} |u|^2 dx \right)^{1/2}$$

or equivalently for $0 < \rho < R$

$$\sup_{B_\rho} u \le \frac{c}{(R-\rho)^{n/2}} \left(\int_{B_R} |u|^2 dx \right)^{1/2}$$

PROOF: Take $x_1 \in B_{\tau R}$, then $u(x_1)^2 > \left(\sup_{B_{\tau R}} |u|\right)^2 - \varepsilon$ for some $\varepsilon > 0$ and

$$\begin{aligned} \left(\sup_{B_{\tau R}} |u|\right)^2 < \varepsilon + u(x_1)^2 &\le \varepsilon + \left(\sup_{B_{(I-\tau)R/4}(x_1)} |u| \right)^2 \\ &\le \varepsilon \frac{c}{|B_{(1-\tau)R/2}(x_1)|} \int_{B_{(I-\tau)R/4}(x_1)} |u|^2 dx \\ &\le \varepsilon + \frac{2^n c}{(1-\tau)^n} \fint_{B_R} |u|^2 dx . \end{aligned}$$

2) *For all $p, 0 < p \le 2$, we have*

$$\sup_{B_\rho} |u| \le \frac{c(p)}{(R-\rho)^{n/p}} \left(\int_{B_R} |u|^p dx \right)^{1/p} \qquad \textit{for all } \rho < R .$$

PROOF: By 1)

$$
\begin{aligned}
\sup_{B_\rho}|u| &\leq \frac{c}{(R-\rho)^{n/2}}\left(\int_{B_R}|u|^2dx\right)^{1/2}\\
&\leq \frac{c}{(R-\rho)^{n/2}}\left(\int_{B_R}|u|^p\left(\sup_{B_R}|u|\right)^{2-p}dx\right)^{1/2}\\
&\leq \frac{c}{(R-\rho)^{n/2}}\left(\int_{B_R}|u|^pdx\right)^{1/2}\left(\sup_{B_R}|u|\right)^{\frac{2-p}{2}}\\
&\leq \varepsilon\sup_{B_R}|u| + \frac{c(\varepsilon)}{(R-\rho)^{n/p}}\left(\int_{B_R}|u|^pdx\right)^{1/p} \qquad \text{for any } \varepsilon>0\ .
\end{aligned}
$$

(We use $ab \leq \varepsilon a^p + \varepsilon^{-q/p}b^q$ for $1/p+1/q=1$, $a\geq 0$, $b\geq 0$.)

Lemma 5.1 *Let $f(t)$ be a nonnegative bounded function on $0 \leq T_0 \leq t \leq T_1$. Suppose that for $T_0 \leq t < s \leq T_1$ we have*

$$f(t) \leq A(s-t)^{-\alpha} + B + \theta f(s)$$

with $\alpha > 0$, $0 \leq \theta < 1$ and A, B nonnegative constants. Then there exists a constant $c = c(\alpha,\theta)$, such that for all $\rho < R$, $T_0 \leq \rho < R \leq T_1$ we have

$$f(\rho) \leq c\left[A(R-\rho)^{-\alpha} + B\right]\ .$$

PROOF: If we define $t_0 := \rho$ and $t_{i+1} := t_i + (1-\tau)\tau^i(R-\rho)(0<\tau<1)$, then we get by induction that

$$f(t_0) \leq \theta^k f(t_k) + \left[\frac{A}{(1-\tau)^\alpha}(R-\rho)^{-\alpha} + B\right]\sum_{i=0}^{k-1}\theta^i\tau^{-i\alpha}\ .$$

Now we choose τ in such a way that $\tau^{-\alpha}\theta < 1$ and if we let $k\to\infty$ we find

$$f(\rho) \leq c(\alpha,\theta)\left[\frac{A}{(R-\rho)^\alpha} + B\right]\ .$$

If we set $T_0 := 0$, $T_1 := R$ and $f(\rho) = \sup_{B_\rho}|u|$, we can immediately conclude from the lemma that

$$\sup_{B_\rho}|u| \leq \frac{c}{(R-\rho)^{n/p}}\left(\int_{B_R}|u|^pdx\right)^{1/p}$$

and in particular for $\rho = R/2$

$$(++) \qquad \sup_{B_{R/2}} |u| \le c \left(\fint_{B_R} |u|^p \right)^{1/p} \qquad 0 < p \le 2 \ .$$

Note that (++) holds for *all* $p > 0$ (of course with $c = c(p)$).

We now come to the main result of this section, namely

De Giorgi's theorem. *If u and $-u$ belong to $DG(\Omega)$, then u is Hölder-continuous, i.e. $u \in C^{0,\alpha}(\Omega)$ for some α. Moreover for $\rho < R < R_0$ we have*

$$\int_{B_\rho} |Du|^2 dx \le c(\rho/R)^{n+2\alpha} \int_{B_R} |u - u_R|^2 dx \ ,$$

or equivalently

$$\int_{B_\rho} |Du|^2 dx \le c(\rho/R)^{n-2+2\alpha} \int_{B_R} |Du|^2 dx$$

We first prove the following

Proposition 5.2 *Set $M(R) := \sup_{x\in B_R} u(x)$ and $m(R) := \inf_{x \in B_R} u(x)$. If we assume that*

$$|A(k_0, R)| \le 1/2|B_R| \quad where \quad k_0 := 1/2[M(2R) + m(2R)]$$

then $\lim_{h \to M(2R)} A(h, R) = 0.$

Proof: For $h > k > k_0$ set $v(x) := \min(u, h) - \min(u, k)$ then

$$\begin{aligned} \operatorname{meas} \{x \in B_R \mid v(x) = 0\} &= |B_R \setminus A(k, R)| \\ &= \operatorname{meas}\{u < k\} \ge \operatorname{meas}\{u < k_0\} \ge 1/2|B_R| \ . \end{aligned}$$

We apply the Sovolev-Poincaré-inequality:

$$\left(\int_{B_R} |v|^{1^*} dx \right)^{1/1^*} \le c \int_{B_R} |Dv| dx \qquad \left(1^* = \frac{n}{n-1}\right) \ ,$$

i.e.

$$\begin{aligned} (h-k)^{n/n-1} |A(h, R)| &\le \int_{A(h,R)} (\min(u, h) - k)^{1^*} dx \\ &\le c \left(\int_{A(k,R)\setminus A(h,R)} |Du| dx \right)^{n/n-1} \end{aligned}$$

and by Hölder's inequality we find

$$\begin{aligned}(h-k)^2|A(h,R)|^{\frac{2n-2}{n}} &\leq c \int\limits_{A(k,R)\setminus A(h,R)} |Du|^2 dx |A(k,R)\setminus A(h,R)| \\ &\leq c \int\limits_{A(k,R)} |Du|^2 dx |A(k,R)\setminus A(h,R)|\end{aligned}$$

and finally, since $u \in DG(\Omega)$,

$$(h-k)^2|A(h,R)|^{\frac{2n-2}{n}} \leq c\, R^{n-2}t[M(2R)-k]^2|A(k,R)\setminus A(h,R)| \ .$$

Now for $M := M(2R)$ we set $k_i := M - \frac{M-k_0}{2^i}$ then

$$k_i - k_{i-1} = \frac{M-k_0}{2^i}$$

and

$$M - k_{i-1} = \frac{M-k_0}{2^{i-1}}$$

hence

$$|A(k_i,R)|^{\frac{2n-2}{n}} \leq 4c\, R^{n-2}\left[|A(k_{i-1},R| - |A(k_i,R)|\right] \ .$$

Summing up for $i = 1, 2, \ldots, \nu$ and using

$$|A(k_i,R)| \geq |A(k_\nu,R)| \qquad \text{for } \ i \leq \nu$$

we find

$$\begin{aligned}\nu\,|A(k_\nu,R)|^{\frac{2n-2}{n}} &\leq 4c\, R^{n-2}\left[|A(k_0,R)| - |A(k_\nu,R)|\right] \\ &\leq 4c\, R^{n-2}\,|A(k_0,R)| \ ;\end{aligned}$$

therefore

$$|A(k_\nu,R)| \leq \left[4c\, R^{n-2}\,|A(k_0,R)|\,\nu^{-1}\right]^{\frac{n}{2n-2}}$$

and

$$\lim_{\nu\to\infty} |A(k_\nu,R)| = 0 \ .$$

Looking at the proof we see that we may also state the following proposition

Proposition 5.3 *Let $u \in H^{1,2}(\Omega)$ and suppose that $|A(k_0,R)| \leq \gamma|B_R|$ for some $\gamma \in (0,1)$; then for all h, k, $h > k > k_0$, we have*

$$(h-k)^2|A(h,R)|^{\frac{2n-2}{n}} \leq \frac{c(\gamma)}{R^2} \int\limits_{A(k,R)} |Du|^2 dx[|A(k,R)| - |A(h,R)|]$$

and if moreover $u \in DG(\Omega)$, then

$$h-k)^2|A(h,R)|^{\frac{2n-2}{n}} \leq \frac{c(\gamma)}{R^2} \int\limits_{B_{2R}} |(u-k)^+|^2\, dx[|A(k,R)| - |A(h,R)|] \ .$$

We now come to the

PROOF OF DE GIORGI'S THEOREM: We may assume that for $k_0 = k_0(u) = 1/2[M(2R) + m(2R)]$ we have $\text{meas}\{x \in B_R \mid u(x) \geq k_0\} > 1/2|B_R|$ (because otherwise $\text{meas}\{x \in B_R \mid -u(x) \geq k_0(-u)\} = \text{meas}\{x \in B_R \mid u(x) \leq k_0(u)\} > 1/2B_R$ and we can work with $-u$). We apply theorem 5.1 with k_0 replaced by $k_\nu := M(2R) - \frac{1}{2^{\nu+1}}[M(2R) - m(2R)]$ and we get

$$M(R/2) \leq k_\nu + \text{const.}\,[M(2R) - k_\nu]\left(\frac{|A(k_\nu, R)|}{R^n}\right)^{\frac{\theta-1}{2}} .$$

Because of the proposition we can choose a ν large enough such that

$$\text{const}\left(\frac{|A(k_\nu, R)|}{R^n}\right)^{\frac{\theta-1}{2}} < 1/2 .$$

Then

$$\begin{aligned} M(R/2) &\leq M(2R) - \frac{1}{2^{\nu+1}}[M(2R) - m(2R)] + \frac{1}{2}[M(2R) - k_\nu] \\ &= M(2R) - \frac{1}{2^{\nu+2}}[M(2R) - m(2R)] . \end{aligned}$$

We now substract $m(R/2)$:

$$\begin{aligned} M(R/2) - m(R/2) &\leq M(2R) - m(R/2) - \frac{1}{2^{\nu+2}}[M(2R) - m(2R)] \\ &\leq [M(2R) - m(2R)] \cdot \left(1 - \frac{1}{2^{\nu+2}}\right) \end{aligned}$$

i.e.

$$\omega(R/2) := M(R/2) - m(R/2) \leq \theta\omega(2R)$$

with $0 < \theta < 1$; thus there exists some constant α such that

$$(**) \qquad \omega(\rho) \leq c(\rho/R)^\alpha \omega(R)$$

and u is Hölder-continuous with exponent α. The estimates in the theorem then follow directly from $(**)$ and the definition of $DG(\Omega)$.

5.2 Moser's iteration technique

Let u be a subsolution for the elliptic operator $-D_\beta(a_{\alpha\beta}D_\alpha)$, i.e.

$$\int_\Omega a^{\alpha\beta}(x) D_\alpha u \, D_\beta \phi dx \leq 0 \qquad \text{for } \forall \phi \in H_0^{1,2}(\Omega) , \quad \phi \geq 0$$

where $a^{\alpha\beta} \in L^\infty(\Omega)$ and satisfy an ellipticity-condition. The *idea* is to use as test-function $\phi := (u^+)^p\eta^2$ with $p > 1$ and η the cut-off-function we defined on page 20. For the sake of simplicity we assume that $u \geq 0$. Then

$$p \int_{B_R} a^{\alpha\beta} D_\alpha u\, D_\beta u\, u^{p-1}\eta^2 dx + \int_{B_R} a^{\alpha\beta} D_\alpha u\, u^p \eta D\eta dx \;\leq\; 0$$

$$\begin{aligned}\int_{B_R} |Du|^2 u^{p-1}\eta^2 dx &\leq \frac{c}{p}\int_{B_R} |Du| u^{\frac{p-1}{2}} u^{\frac{p+1}{2}} \eta |D\eta| dx \\ \int_{B_R} |Du|^2 u^{p-1}\eta^2 dx &\leq \frac{c}{p^2}\int_{B_R} u^{p+1}|D\eta|^2 dx\ .\end{aligned}$$

As

$$|Du^{\frac{p+1}{2}}|^2 = \left(\frac{p+1}{2}\right)^2 u^{p-1}|Du|^2$$

we get

$$\int_{B_R} |Du^{\frac{p+1}{2}}|^2\eta^2 dx \leq c\left(\frac{p+1}{p}\right)^2 \int_{B_R} u^{p+1}|D\eta|^2 dx$$

and together with

$$\left|D(\eta u^{\frac{p+1}{2}})\right|^2 dx \leq c|Du^{\frac{p+1}{2}}|^2\eta^2 + u^{p+1}|D\eta|^2$$

follows

$$\begin{aligned}\int_{B_R} |Du^{\frac{p+1}{2}}|^2 dx &\leq c\left(1+\left(\frac{p+1}{2}\right)^2\right)\int_{B_R} u^{p+1}|D\eta|^2 dx \qquad (p>1)\\ &\leq c(p+1)^2\left(1+\frac{1}{p^2}\right)\int_{B_R} u^{p+1}|D\eta|^2 dx\ .\end{aligned}$$

By Sobolev's inequality and the properties of η (cf. p. 20) we finally have:

$$\left(\int_{B_R} (u^{\frac{p+1}{2}}\eta)^{2^*} dx\right)^{2/2^*} \leq c(p+1)^2\left(1+\frac{1}{p^2}\right)\frac{1}{(R-\rho)^2}\int_{B_R} u^{p+1} dx$$

For $p = 1$ this is just Caccioppoli's inequality. Set $\lambda := 2^*/2 = \frac{n}{n-2}$ and $q := p+1 > 2$ then

$$\left(\int_{B_\rho} u^{\lambda q} dx\right)^{1/\lambda} \leq c\frac{(1+q)^2}{(R-\rho)^2}\int_{B_R} u^q dx\ .$$

Now we choose

$$\begin{array}{rcll} q_i & := & 2\lambda^i = \lambda q_{i-1} & (q_0 = 2) \\ R_i & := & \frac{R}{2} + \frac{R}{2^{i+1}} & (R_0 = R) \end{array}$$

and we find that

$$\begin{aligned} \left(\int\limits_{B_{R_{i+1}}} u^{q_{i+1}} dx \right)^{\frac{1}{\lambda^{i+1}}} &= \left(\int\limits_{B_{R_{i+1}}} u^{\lambda q_i} dx \right)^{\frac{1}{\lambda}\frac{1}{\lambda^i}} \\ &\leq \left[\frac{c(1+q_i)^2}{2^{-2(i+1)} R^2} \right]^{1/\lambda^i} \left(\int\limits_{B_{R_i}} u^{q_i} dx \right)^{1/\lambda^i} \\ &\leq \prod_{k=0}^{i} \left[\frac{c(1+q_k)^2}{R^2 4^{-k-1}} \right]^{1/\lambda^k} \left(\int\limits_{B_R} u^2 dx \right)^{1/2} . \end{aligned}$$

We have to estimate the above product:

$$\begin{aligned} \prod_{k=0}^{\infty} \left[\frac{c(1+q_k)^2}{R^2 4^{-k-1}} \right]^{1/\lambda^k} &= \exp\left(\log \prod_{k=0}^{\infty} \left[\frac{c(1+q_k)^2}{R^2} 4^{-k-1} \right]^{1/\lambda^k} \right) \\ &= \exp \sum_{k=0}^{\infty} \frac{2}{\lambda^k} \log \hat{c} \frac{(1+q_k)}{R 2^{-k-1}} \\ &= \exp(\bar{c} - n \log R) \left(q_k = 2\lambda^k, \sum_{j=0}^{\infty} \lambda^{-j} = \frac{n}{2} \right) \end{aligned}$$

thus for $\forall k$

$$\left(\fint\limits_{B_{R_k}} u^{q_k} dx \right)^{1/q_k} \leq \tilde{c} \left(\fint\limits_{B_R} u^2 dx \right)^{1/2}$$

from which immediately follows that

$$\sup_{x \in B_{R/2}} u(x) \leq \tilde{c} \left(\fint\limits_{B_R} |u|^2 dx \right)^{1/2} .$$

Remark: Instead of 2 one can take any exponent $p > 0$ in the above formulas.

If we start with a supersolution and $u > 0$, i.e.

$$\int\limits_{\Omega} a^{\alpha\beta}(x) D_\alpha u \, D_\beta \phi dx \geq 0 \qquad \text{for } \forall \phi \in H_0^{1,2}(\Omega) \, , \quad \phi \geq 0$$

then we can take $p < -1$.

Exactly as before we then get for $\lambda := \frac{2^*}{2} > 1$ and $q := p + 1 < 0$:

$$\left(\int_{B_\rho} u^{\lambda q} dx\right)^{1/\lambda} \leq c(1+q^2)\left(1 + \frac{1}{(q-1)^2}\right)\frac{1}{(R-\rho)^2}\int_{B_R} u^q dx \ .$$

As $|q-1| > 1$ we have

$$\left(\int_{B_\rho} u^{\lambda q} dx\right)^{1/\lambda} \leq c\frac{(1+q)^2}{(R-\rho)^2}\int_{B_R} u^q dx \ .$$

Now we set $R_i := \frac{R}{2} + \frac{R}{2^{i+1}}$ as before and $q_i := q_0\lambda^i$ but $q_0 < 0$. Then follows

$$\inf_{x \in B_{R/2}} u(x) \leq c\left(\fint_{B_R} u^{q_0} dx\right)^{1/q_0} \qquad (q_0 < 0) \ .$$

Up to now we have the following two *estimates for a positive solution* u: For any $p > 0$ and any $q < 0$ holds:

$$\sup_{x \in B_{R/2}} u(x) \ \leq \ k_1 \left(\fint_{B_R} u^p dx\right)^{1/p}$$

$$\inf_{x \in B_{R/2}} u(x) \ \geq \ k_2 \left(\fint_{B_R} u^q dx\right)^{1/q} \qquad \text{for some constants } k_1 \text{ and } k_2,\ q < 0.$$

The main point now is that the John-Nirenberg-lemma allows us to combine the two inequalities to get

Harnack's inequality

If u is a solution and $u > 0$, then

$$\inf_{x \in B_{R/2}} u(x) \geq c \sup_{x \in B_{B/2}} u(x) \ .$$

PROOF: We have

$$\int_\Omega a^{\alpha\beta} D_\alpha u D_\beta \phi dx = 0 \qquad \text{for } \forall \phi \in H_0^{1,2}(\Omega) \qquad \phi \geq 0$$

As $u > 0$ we can choose $\phi := \frac{1}{u}\eta^2$ and get

$$-\int\limits_\Omega a^{\alpha\beta} D_\alpha u \, D_\beta u \frac{1}{u^2}\eta^2 + \int\limits_\Omega a^{\alpha\beta} D_\alpha u \frac{1}{u}\eta D\eta = 0$$

thus $\int\limits_\Omega \frac{|Du|^2}{u^2}\eta^2 dx \le \int\limits_\Omega |D\eta|^2 dx$, i.e. $\int\limits_{B_R} |D\log u|^2 dx \le c\, R^{n-2}$ which implies by Poincaré's inequality that $\log u \in BMO$.

Now we use the characterization of BMO ((4) p. 68) by which there exists a constant α, such that for $v := e^{\alpha \log u}$ we have

$$\int\limits_{B_R} v dx \le c \left(\int\limits_{B_R} \frac{1}{v} dx \right)^{-1} .$$

Hence $(p = 1, q = -1)$

$$\inf_{B_{R/2}} u \ge k_2 \left(\fint_{B_R} u^{-a} dx \right)^{-1/\alpha} \ge k_2 c^{-1} \left(\fint_{B_R} u^{\alpha} dx \right)^{1/\alpha} \ge \frac{K_2}{k_1 c} \sup_{B_{R/2}} u \ .$$

Remarks:

1) It follows from Harnack's inequality that a solution of the equation at the beginning of this section is Hölder-continuous:
$M(2R) - u \ge 0$ is a supersolution, thus

$$M(2R) - m(R) \le c[M(2R) - M(R)]$$

$u - m(2(R)) \ge 0$ is also a supersolution, thus

$$M(R) - m(2R) \le c[m(R) - m(2R)]$$

and from this we get

$$\omega(2R) + \omega(R) \le c[\omega(2R) - \omega(R)]$$

hence $\omega(R) \le \frac{c-1}{c+1}\omega(2R)$ and u is Hölder-continuous.

2) Sometimes one works with other domains B_r than balls; for example in case of parabolic equations one uses "parabolic cubes", in the theory of minimal surfaces one uses the intersection of balls (or cylinders) in R^{n+1} with an n-dimensional minimal surface.

Then the estimates at the beginning of p. 87 still hold, but is is difficult (and sometimes not possible) to extend the John-Nirenberg-theorem in order to accomplish the proof of Harnack's inequality as before.

Nevertheless this can be done using the following

Abstract John-Nirenberg-theorem (Bombieri-Giusti). *Let S be a topological space, μ a Borel-regular positive measure on S and $\{B_r\}$ a family of open sets with $B_s \subset B_r$ for $s \leq r$ and $0 < \mu(B_r) < \infty$ for $\forall r \in (0,1)$.*
For $p \neq 0$ we set

$$\begin{aligned} |u|^*_{p,r} &:= \left(\frac{1}{\mu(B_r)} \int_{B_r} u^p d\mu \right)^{1/p} \\ |u|^*_{\infty,r} &:= \sup_{B_r} u \\ |u|^*_{-\infty,r} &:= \inf_{B_r} u \ . \end{aligned}$$

Now for $\theta_0, \theta_1 > 0$ (∞ is allowed) $u \geq 0$ suppose $|u|^*_{\theta_0,1} < \infty$ and $|u|^*_{-\theta_1,1} > 0$ and moreover that there exist $\sigma > 0$, $0 < p_0 < 1/2 \min(\theta_0, \theta_1)$, $0 < Q < 1/3$ such that:

For $\forall s, r, 0 < s < r \leq 1$ and $\forall p, 0 < p < p_0$

$$\begin{aligned} |u|^*_{\theta_0,s} &\leq \{Q(r-s)^\sigma\}^{\frac{1}{\theta_0} - \frac{1}{p}} |u|^*_{p,r} \\ |u|^*_{-\theta_1,s} &\geq \{Q(r-s)^\sigma\}^{\frac{1}{p} - \frac{1}{\theta_1}} |u|^*_{-p,r} \end{aligned}$$

moreover

$$A := \sup_{0 \leq r \leq 1} \inf_{\lambda > 0} \left\{ \frac{1}{\mu(B_r)} \int_{B_r} \left| \log \frac{u}{\lambda} \right| d\mu \right\} < \infty$$

then

$$|u|^*_{\theta_0,0} \leq \left[\frac{\mu(B_1)}{\mu(B_0)} \right]^{\frac{1}{\theta_0} \frac{1}{\theta_1}} \exp\left\{ c_1 Q^{-2} \left(\frac{1}{p_0} + A \right) \right\} |u|^*_{-\theta_1,0}$$

where $c_1 = c_1(\sigma)$.

5.3 De Giorgi-class and Harnack-inequality

The following theorem relates the De Giorgi class to Harnack's inequality.

Theorem 5.5 (Di Benedetto-Trudinger) *If* $-u \in DG(\Omega)$, $u > 0$ *then*

$$\inf_{B_R} u \geq c \left(\fint_{B_{2R}} |u|^p dx \right)^{1/p}$$

for some $p > 0$ (weak Harnack-inequality).

PROOF:

Step 1: As $-u \in DG(\Omega)$ we can apply the theorem p. 79:

Choose $k_0 = -\tau$, $\tau > 0$ then

$$\sup_{B_{R/2}} (-u) \leq -\tau + c \left(\frac{1}{R^n} \int_{B_R \cap \{-u > -\tau\}} (-u + \tau)^2 dx \right)^{1/2} \cdot \left(\frac{|B_R \cap \{-u > -\tau\}|}{|B_R|} \right)^{\frac{\theta-1}{2}}$$

and

$$\begin{aligned} \inf_{B_{R/2}} u &\geq \tau - c \left(\frac{1}{|B_R|} \int_{B_R \cap \{u < \tau\}} (\tau - u)^2 dx \right)^{1/2} \left(\frac{|B_R \cap \{u < \tau\}|}{|B_R|} \right)^{\frac{\theta-1}{2}} \\ &\geq \tau - c\tau \left(\frac{|B_R \cap \{u < \tau\}|}{|B_R|} \right)^{\theta/2} \end{aligned}$$

first result:

If $\text{meas}\{x \in B_R | u(x) < \tau\} \leq \gamma_0 |B_R|$ *with*

$$\gamma_0 = \left(\frac{1}{2c} \right)^{2/\theta}$$

then

$$\inf_{B_{R/2}} u \geq 1/2\tau \ .$$

Step 2: Suppose only that $\text{meas}\{x \in B_R | u(x) < \tau\} \leq \gamma |B_R|$ for some $\gamma \in (0,1)$, i.e.

$$\text{meas}\{x \in B_R \mid -u(x) > -\tau\} \leq \gamma |B_R| \ .$$

We apply proposition p. 79 with $h = k_{s+1}$ and $k = k_s = -\frac{\tau}{2^s}$ ($k_0 = -\tau$) then

$$\underbrace{(k_{s+1} - k_s)^2}_{\tau^2 4^{-s-1}} \operatorname{meas}\left\{x \in B_R \mid -u > -\frac{\tau}{2^{s+1}}\right\}^{\frac{2n-2}{n}} \leq$$
$$\leq \frac{c(\gamma)}{R^2} \underbrace{\int_{B_{2R}} (k_s - u)^2 dx}_{\leq \frac{\tau^2}{4^s} R^n} \left[\left|\left\{-u > -\frac{\tau}{2^s}\right\}\right| - \left|\left\{-u > -\frac{\tau}{2^{s+1}}\right\}\right| \right]$$

and hence

$$\operatorname{meas}\left\{x \in B_R \mid u < \frac{\tau}{2^{s+1}}\right\}^{\frac{2n-2}{n}} \leq$$
$$\leq 4c(\gamma) R^{n-2} \left[\operatorname{meas}\left\{x \in B_R \mid u < \frac{\tau}{2^s}\right\} - \operatorname{meas}\left\{x \in B_R \mid u < \frac{\tau}{2^{s+1}}\right\}\right]$$

summing up from $s = 0$ to $s = \nu$ we compute

$$\begin{aligned} \nu \operatorname{meas}\left\{x \in B_R \mid u < \frac{\tau}{2^{\nu+1}}\right\}^{\frac{2n-2}{n}} &\leq 4c(\gamma) R^{n-2} \operatorname{meas}\{x \in B_R \mid u < \tau\} \\ &\leq 4c(\gamma) R^{n-2} \gamma |B_R| \\ &= \tilde{c}(\gamma) \gamma |B_R|^{\frac{n-2}{n}} |B_R| \ , \end{aligned}$$

i.e. we have the

second result:

If $\operatorname{meas}\{x \in B_R \mid u < \tau\} \leq \gamma |B_R|$ *for* $\gamma \in (0,1)$ *then*

$$\operatorname{meas}\left\{x \in B_R \mid < \frac{\tau}{2^{\nu+1}}\right\} \leq \left(\frac{\tilde{c}(\gamma)\gamma}{\nu}\right)^{\frac{n}{2n-2}} |B_R| \ .$$

If we fix ν in such a way that

$$\left(\frac{\tilde{c}(\gamma)\gamma}{\nu}\right)^{\frac{n}{2n-2}} \leq \gamma_0 \ ,$$

we immediately deduce from step 1 that

$$\inf_{B_{R/2}} u \geq \frac{\tau}{2^{\nu+2}} \ .$$

Taking everything together we have shown: *If for some* $\delta \in (0,1)$

$$\operatorname{meas}\{x \in B_R \mid u(x) < \tau\} \leq \delta |B_R|$$

then $\inf_{B_{R/2}} u \geq c(\delta)\tau$.

Now we observe that for some δ fixed we have:

If $\text{meas}\{x \in B_R \mid u \geq \tau\} \geq \delta|B_R| = \frac{\delta}{6^n}|B_{6R}|$ then $\text{meas}\{x \in B_{6R} \mid u \geq \tau\} \geq \text{meas}\{x \in B_R \mid u \geq \tau\} \geq \frac{\delta}{6^n}|B_{6R}|$ and thus $\inf\limits_{B_{3R}} \geq c(\delta)\tau$. So we have actually proved:

If $\text{meas}\{x \in B_R \mid u < \tau\} \geq \gamma|B_R|$ *then* $\inf\limits_{B_{3R}} u \geq \lambda\tau$ *where* $\lambda = \lambda(\gamma)$.

Step 3 (Krylov-Safanov-covering argument):

Let Q_R be a cube in $\mathbb{R}^n$, E a measurable set in Q_R and $\delta \in (0,1)$ a fixed number. We look to the set

$$E_\delta := \bigcup_{\substack{x \in Q_R \\ \rho > 0}} \{Q_{3\rho}(x) \cap Q_R \mid |E \cap Q_{3\rho}(x)| \geq \delta Q_\rho\}$$

then either $E_\delta = Q_R$ or $|E_\delta| \geq \delta^{-1}|E|$.

Proof: If $|E| \geq \delta|Q_R|$ it is obvious (take $\rho = R$) that $E_\delta = Q_R$. Consequently we may assume that $|E| < \delta|Q_R|$. We subdivide Q_R as in the Calderon-Zygmund argument into 2^n equal subcubes and subdivide again the cubes for which $|E \cap Q| < \delta|Q|$.

Let $\mathcal{S}$ be the family of subcubes of Q such that $|E \cap Q| \geq \delta|Q|$. We define $\tilde{E}_\delta := \bigcup\limits_{k \in \mathcal{S}} \tilde{K}$, $\tilde{K}$ being the last cube for which $|E \cap \tilde{K}| < \delta|\tilde{K}|$ and K is one of the equal subcubes of the subdivision of $\tilde{K}$.

Clearly $\tilde{E}_\delta \subset E_\delta$ and therefore

$$|\tilde{E}_\delta \cap E| = \sum_{K \in \mathcal{S}} |\tilde{K} \cap E| \leq \delta \sum_{K \in \mathcal{S}} |\tilde{K}| = \delta|\tilde{E}_\delta| \leq \delta|E_\delta| \ .$$

Finally we have also $|\tilde{E}_\delta \cap E| = |E|$.

We apply this to the set $E = \{x \in Q_R \mid u(x) > t\lambda^{i-1}\} =: A_t^{(i-1)} (\subset A_t^i \forall i)$.

If for some $z \in Q_R$ and some $\rho > 0$ we have $|E \cap Q_{3\rho}(z)| \geq \delta|Q_\rho(z)|$ then $u(x) \geq \lambda^i t$ for $\forall x \in Q_{3\rho}(z)$ (cf. p. 91) hence

$$\text{meas}\left\{x \in Q_R \mid u \geq \lambda^i t\right\} \geq \frac{1}{\delta}\text{meas}\left\{x \in Q_R \mid u(x) > \lambda^{i-1} t\right\}$$

or

$$\text{meas}\left\{x \in Q_R \mid u \geq \lambda^i t\right\} = |Q_R|$$

by the Krylov-Safanov argument. In both cases we conclude:

If $|A_t^{(0)}| > \delta^s |Q_R|$ for some s, then

$$|A_t^{(s-1)}| \geq \frac{1}{\delta}|A_t^{(s-2)}| \geq \ldots \geq \delta^{1-s}|A_t^{(0)}| \geq \delta |Q_R| \ .$$

To finish the proof we choose for $\forall t > 0$ s in such a way that

$$\delta^s |Q_R| \leq |A_t| = \text{meas}\{x \in Q_R \mid u > t\}$$

i.e. $s \geq \log \frac{|A_t|}{|Q_R|} / \log \delta$. This implies

$$\inf_{Q_{3R}} u \geq \frac{1}{2^s} t = \exp\left(\frac{\log 2}{\log \delta} \log \frac{|A_t|}{|Q_R|}\right) t = c_1 t \left(\frac{|A_t|}{|Q_R|}\right)^{c_0}$$

i.e.

$$\frac{|A_t|}{|Q_R|} \leq c_2 t^{-1/c_0} \left(\inf_{Q_{3R}} u\right)^{1/c_0} \quad \left(c_2 = \left(\frac{1}{c_1}\right)^{1/c_0}\right)$$

which means that

$$u \in L_w^{1/c_0}(\Omega)!$$

Now as

$$\fint_{Q_R} u^p dx = \frac{1}{|Q_R|} \left\{ p \int_\xi^\infty t^{p-1} |A_t| dt + |A_\xi| \xi^p \right\}$$

we can choose $\xi = \inf\limits_{Q_{3R}} u$ to get

$$\begin{aligned}
\fint_{Q_R} u^p dx &\leq p \int_\xi^\infty t^{p-1} c_2 t^{-1/c_0} \left(\inf_{Q_{3R}} u\right)^{1/c_0} dt \\
&= p c_2 \left(\inf_{Q_{3R}} u\right)^{1/c_0} \int_\xi^\infty t^{p-1-1/c_0} dt \\
&= \frac{c_2 p}{\left(p - \frac{1}{c_0}\right)} \left(\inf_{Q_{3R}} u\right)^{1/c_0} t^{p-1/c_0} \Bigg|_{t=\xi}^{t=\infty} \\
&= \frac{c_2 p}{\frac{1}{c_0} - p} \left(\inf_{Q_{3R}} u\right)^p \qquad \text{provided } p < \frac{1}{c_0}
\end{aligned}$$

in conclusion we have proved

$$\inf_{Q_{R/2}} u \geq \inf_{Q_{3R}} u \geq c \left(\fint_{Q_R} u^p dx\right)^{1/p} \qquad \text{for some positive } p$$

(of course it is not relevant if we have Q_R or B_R), and the theorem is proved.

5.4 Quasi-minima

If we want to minimize a quadratic functional $\int_\Omega F(Du)dx$ then, by the above result, we have $Du \in C^{0,\alpha}(\Omega)$ for a minimizer u. The situation completely changes if we want to minimize a functional of the form

$$(*) \qquad \int_\Omega F(x,u,Du)dx =: \mathbb{F}(u;\Omega) \ .$$

One still can compute the Euler-equation

$$\int_\Omega [F_p(x,u,Du)D\phi + F_u(x,u,Du)\phi]\, dx = 0 \qquad \text{for } \forall\phi \in C_0^\infty(\Omega) \ ,$$

but it is no more possible to take as test function $\phi = D_s\psi$!

The idea to deal with $(*)$ is to show that a minimizer u is in $DG(\Omega)$, see Giaquinta-Giusti [26].

For the sake of simplicity let us assume the growth condition

$$|p|^2 \le F(x,u,p) \le c|p|^2 \ .$$

If

$$(**) \qquad \mathbb{F}(u, \operatorname{supp}\phi) \le \mathbb{F}(u+\phi; \operatorname{supp}\phi)$$

for all $\phi \in C_0^\infty$ we call u a *minimizer.*

If $(**)$ holds for all $\phi \le 0$ (resp. all $\phi \ge 0$) we call u a *subminimizer* (resp. *superminimizer*). If u is a minimizer we have

$$\int_{B_R \cap \operatorname{supp}(u-v)} F(x,u,Du)dx \le \int_{B_R \cap \operatorname{supp}(u-v)} F(x,v,Dv)dx$$

for all $v - u \in H_0^{1,2}$.

Choose $v = u - \eta\max(u-k,0)$ where $\eta = 1$ on B_ρ and $\eta \in C_0(B_R)$. We then have

$$\begin{aligned}
\int_{A(k,\rho)} |Du|^2 dx &\le \int_{B_R \cap \operatorname{supp}(u-v)} |Du|^2 dx \\
&\le c \int_{B_R \cap \operatorname{supp}(u-v)} |D(u - \eta\max(u-k,0))|^2 dx \\
&\le c \int_{A(k,R)} (1-\eta)^2 |Du|^2 dx \frac{c}{(R-\rho)^2} \int_{A(k,R)} |u-k|^2 dx \\
&\le c \int_{A(k,R)-A(k,\rho)} |Du|^2 dx + \frac{c}{(R-\rho)^2} \int_{A(k,R)} |u-k|^2 dx \ .
\end{aligned}$$

On both sides we add $\int\limits_{A(k,\rho)} |Du|^2 dx$ to get

$$(1+c) \int\limits_{A(k,\rho)} |Du|^2 dx \leq c \int\limits_{A(k,R)} |Du|^2 dx + \frac{c}{(R-\rho)^2} \int\limits_{A(k,R)} |u-k|^2 dx \ .$$

We divide by $(1+c)$ and apply lemma (5.1) so that we finally have

$$\int\limits_{A(k,\rho)} |Du|^2 dx \leq \frac{c}{(R-\rho)^2} \int\limits_{A(k,R)} |u-k|^2 dx \ , \qquad \text{i.e. } u \in DG(\Omega).$$

Thus u is Hölder-continuous and, if in addition $u > 0$, it satisfies Harnack's inequality.

More generally let us introduce the notion Q-minima: We say that $u \in H^{1,2}_{\text{loc}}(\Omega)$ is a *quasiminima for the functional* $(*)$ *with constant* Q, if

(+) $\qquad \mathbb{F}(u; \operatorname{supp} \phi) \leq Q\mathbb{F}(u+\phi; \operatorname{supp} \phi)$ for all ϕ with supp $\phi c \subset \Omega$.

In a similar way we define *super-Q-minima* and *sub-Q-minima.* Then the argument above leads to the conclusion:

If u is a Q-minimum for $\mathbb{F}$, then u is Hölder continuous.

If u is a positive Q-minimum, then u satisfies Harnack's inequality.

We shall not discuss the notion of Q-minima further and refer to [30], [22] for details. Here we only want to note that

a) minimizers of $\mathbb{F}$ are of course Q-minimizers for $\mathbb{F}$ and more precisely quasi-minima for the Dirichlet-integral,

b) solution of elliptic equations

$$\int\limits_{\Omega} A^{\alpha\beta}(x) D_\alpha u \, D_\beta \phi dx = 0 \qquad \text{for } \forall \phi \in H^{1,2}_0(\Omega)$$

are quasi-minima for the Dirichlet-integral.

In order to see this, it is sufficient to choose $\phi = u + \psi$ and use the ellipticity and Hölder's inequality.

That already shows the unifying character of this notion and points out that regularity in the scalar case actually relies on a minimality condition (namely (+)). Actually this notion includes many other things for which we also refer to [30], [22].

5.5 Hölder-continuity of derivatives of minimizers

We look at a minimizer of the functional

$$(*) \qquad \mathbb{F}(u;\Omega) = \int_\Omega F(x,u,Du)dx$$

for which we assume

(i) $c|p|^2 \le F(x,u,p) \le |p|^2$

(ii) F is of class C^2 with respect to p and
$|F_{pp}(x,u,p)| \le L = \text{constant}$
$F_{p_\alpha p_\beta}(x,u,p)\xi_\alpha\xi_\beta \ge \nu|\xi|^2$ for $\forall \xi \in \mathbb{R}^n$, $\nu > 0$.

(iii) $|p|^{-2}F(x,u,p)$ is continuous in (x,u) uniformly in p (i.e. there exists a bounded, continuous, concave and increasing function $\omega(t)$ with $\omega(0) = 0$ and
$|F(x,u,p) - F(y,v,p)| \le |p|^2\omega(|x-y|^2 + |u-v|^2)$.

We already know from section 5.4, that any minimizer of $(*)$ is Hölder-continuous with some positive exponent α. With the above assumptions we can prove the following

Theorem 5.6 *Let u be a minimizer of $(*)$. Then u is Hölder-continuous with all exponents $\alpha \in (0,1)$. More precisely: Du belongs to the Morrey-space $L^{2,n-\varepsilon}_{\text{loc}}(\Omega)$ for every $\varepsilon > 0$, and for all $B\rho \subset B_R \subset \Omega$ we have*

$$\int_{B_\rho} |Du|^2 dx \le c(\rho/R)^{n-\varepsilon} \int_{B_R} |Du|^2 dx$$

where $c = c(\varepsilon)$.

PROOF: For $B_R(x_0) \subset \Omega$ we define $F^0(p) := F(x_0, u_{x_0R}, p)$ and consider the problem

$$\mathbb{F}^0(v) = \int_{B_R} F^0(Dv)dx \longrightarrow \text{minimize}$$

$v = u$ on ∂B_R (i.e. $v - u \in H^{1,2}_0(B_R/x_0)))$.

Step 1: We estimate the oscillation of v by the oscillation of u:
For $k > k_0 := \sup_{B_R} u$ we have $\mathbb{F}^0(v;B_R) \le \mathbb{F}^0(\min(v,k);B_R)$ and thus $\sup_{B_R} v \le k$ for $\forall k > k_0$, hence $\sup_{B_R} v \le k_0 = \sup_{B_R} u$.
With a similar argument $\inf_{B_R} v \ge \inf_{B_R} u$

Step 2: We estimate $\sup_{B_{R/2}} |Dv|$. We know that $v \in H^{2,2}_{\text{loc}}(B_R)$ (cf. 2.4) and that any $D_s v$ satisfies

$$\int_{B_R} F^0_{p_\alpha p_\beta}(Dv) D_\beta(D_s v) D_\alpha \phi dx = 0 \qquad \text{for } \forall \phi \in H^{1,2}_0(B_R) \ .$$

We choose $\phi = \eta D_s v$, $\eta \geq 0$, and sum over s, then we have

$$\frac{1}{2} \int_{B_R} F^0_{p_\alpha p_\beta} D_\alpha \left(|Dv|^2\right) D_\beta \eta dx + \int_{B_R} F^0_{p_\alpha p_\beta} D_\alpha(D_s v) D_\beta(D_s v) \eta dx = 0 \ ;$$

the second term is nonnegative by ellipticity (ii) and therefore

$$\int_{B_R} F^0_{p_\alpha p_\beta} D_\alpha \left(|Dv|^2\right) d_\beta \eta dx \leq 0 \qquad \text{for all } \eta \in C^\infty_0(B_R) \ , \ \eta \geq 0$$

thus

$$\sup_{B_{R/2}} |Dv|^2 \leq c \fint_{B_R} |Dv|^2 \qquad \text{(see De Giorgi's theorem 5.1).}$$

In conclusion we have shown

$$\int_{B_\rho} |Dv|^2 dx \leq c(\rho/R)^n \int_{B_R} |Dv|^2 dx \ ,$$

which is trivial for $R/2 < \rho < R$ and for $0 < \rho < R/2$ follows from $\int_{B_\rho} |Dv|^2 dx \leq c\rho^n \sup_{B_{R/2}} |Dv|^2$ and step 2.

Step 3: We come back to u:

$$(**) \quad \begin{aligned} \int_{B_\rho} |Du|^2 dx &\leq 2 \int_{B_\rho} |Dv|^2 dx + 2 \int_{B_R} |D(u-v)|^2 dx \\ &\leq c(\rho/R)^n \int_{B_R} |Du|^2 dx + c \int_{B_R} |D(u-v)|^2 dx \ . \end{aligned}$$

In order to estimate the last term we compute (F^0 is convex!)

$$\begin{aligned} &F^0(Du) - F^0(Dv) \\ &\leq F^0_{p_\alpha}(Dv) D_\alpha(u-v) \\ &\quad + \int_0^1 (1-t) F^0_{p_\alpha p_\beta}(t\, Du + (1-t)Dv) dt\, D_\alpha(u-v) D_\beta(u-v) \\ &\leq F^0_{p_\alpha}(Dv) D_\alpha(u-v) + 1/2\nu |D(u-v)|^2 \ . \end{aligned}$$

Now we integrate over B_R and get (using the Euler-equation

$$0 = \int_{B_R} F^0_{p_\alpha}(Dv) D_\alpha(u-v)dx)$$

that

$$\int_{B_R} |D(u-v)|^2 dx \leq \frac{2}{\nu} \int_{B_R} \left[F^0(Du) - F^0(Dv)\right] dx \ .$$

Now

$$\begin{aligned}
&\int_{B_R} [F(x_0, u_{x_0,R}, Du) - F(x_0, u_{x_0,R} Dv)]\, dx \\
&= \int_{B_R} \Big[F(x_0, u_{x_0,R}, Du) - F(x, u(x), Du) + \\
&+ \underbrace{F(x,u,Du) - F(x,v,Dv)}_{\leq 0 \text{ as } u \text{ is a minimizer with the same boundary values a } v)} + \\
&+ F(x,v,Dv) - F(x_0, u_{x_0,R}, Dv)\Big] \\
&\leq \int_{B_R} \omega\left(|x-x_0|^2 + |u-u_{x_0,R}|^2\right) |Du|^2 dx \\
&+ \int_{B_R} \omega\left(|x-x_0|^2 + |v-v_{x_0,R}|^2\right) |Dv|^2 dx \\
&\leq c\omega(R) \left[\int_{B_R} |Du|^2 dx + \int_{B_R} |Dv|^2 dx\right] ,
\end{aligned}$$

where we used that u is Hölder-continuous and that for $y \in \partial B_R$ one has

$$|v - u_{x_0,R}| \leq |v - v(y)| + |u(y) - u_{x_0,R}| \ .$$

Thus we have

$$\int_{B_R} |D(u-v)|^2 dx \leq c\omega(R) \int_{B_R} |Du|^2 dx \ ;$$

we insert this into $(**)$ and get

$$\begin{aligned}
\int_{B_\rho} |Du|^2 dx &\leq c(\rho/R)^n \int_{B_\rho} |Du|^2 dx + \tilde{c}\omega(R) \int_{B_R} |Du|^2 dx \\
&\leq \bar{c}\left[(\rho/R)^n + \omega(R)\right] \int_{B_R} |Du|^2 dx \ .
\end{aligned}$$

To finish the proof we just have to apply the lemma p. 44 for $\varepsilon = \omega(R)$ and $B = 0$, then

$$\int_{B_\rho} |Du|^2 dx \le \hat{c}(\rho/R)^n \int_{B_R} |Du|^2 dx \ .$$

If we have instead of (iii) an assumption

(iii)′ $|p|^{-2}F(x,u,p)$ is Hölder-continuous in (x,u) uniformly with respect to p, i.e. $\omega(t) \le At^\sigma$ for some constant $A > 0$ and $\xi > 0$, we have the

Theorem 5.7 *Any minimizer u has Hölder continuous first derivatives.*

PROOF: Let $B_R \subset \Omega$ and v as before. Then Dv satisfies

$$\int_{B_R} F^0_{p_\alpha p_\beta}(Dv) D_\beta(D_s v) D_\alpha \phi dx = 0 \qquad \text{for } \forall \phi \in H^{1,2}_0(B_R)$$

and from De Giorgi's theorem we know that

$$\int_{B_\rho} |Dv - (Dv)_\rho|^2 \, dx \le c(\rho/R)^{n+2\delta} \int_{B_R} |Dv - (Dv)_R|^2 \, dx$$

for some positive δ.

Then

$$\int_{B_\rho} |Du - (Du)_\rho|^2 \, dx \le c(\rho/R)^{n+2\delta} \int_{B_R} |Du - (Du)_R|^2 \, dx + c \int_{B_R} |D(u-v)|^2 dx$$

and then apply the lemma p. 44 to conclude that $Du \in C^{0,\mu}(\Omega)$ for some μ.

Remark: The results of this section can be extended up to boundary cf. [32]. Moreover one has optimal Hölder-exponents for the first derivatives cf. [33].

Chapter 6

Regularity in the vector-valued case

When we pass from the regularity theory for scalar minimizers or solutions of elliptic equations to the regularity theory for vector-valued minimizers or solutions of elliptic systems the situation completely changes: *Except in dimension* $n = 2$ *everywhere regularity is a rare phenomena.*

In 1968 De Giorgi [12] showed that his regularity result for solutions of second order elliptic equations with measurable bounded coefficients cannot be extended to solutions of elliptic systems. He presented a quadratic functional

$$\text{(i)} \qquad \int_\Omega A_{ij}^{\alpha\beta}(x) D_\alpha u^i D_\beta u^j \, dx$$

with $A_{ij}^{\alpha\beta} \in L^\infty(\Omega)$ and elliptic, i.e.

$$\text{(ii)} \qquad A_{ij}^{\alpha\beta} \xi_\alpha^i \xi_\beta^j \geq \nu |\xi|^2 \ \forall \xi \in \mathbb{R}^{nN}, \ \nu > 0,$$

having as minimizer an unbounded function in $H^{1,2}(\Omega, \mathbb{R}^N)$. More recently J. Souček showed that minimizers of functionals of the type (i), (ii) may be discontinuous on a dense subset of Ω, see [68], [22].

Modifying De Giorgi's example Giusti-Miranda [40] showed that solutions of elliptic quasilinear systems of the type

$$\int_\Omega A_{ij}^{\alpha\beta}(u) D_\alpha u^i D_\beta \phi^j \, dx = 0 \qquad \text{for } \forall \phi \in C_0^\infty(\Omega, \mathbb{R}^N)$$

with analytic coefficients $A_{ij}^{\alpha\beta}$ satisfying (ii) have singularities in dimension $n \geq 3$.

Similar examples were presented in the meantime independently by Maz'ya [50].

In 1975 J. Nečas presented an elliptic functional of the simplest type

$$\int_\Omega F(Du)dx$$

having as minimizer the vector valued function $(u_{ij}(=\left(\frac{x_i x_j}{|x|}\right)$, which is Lipschitz-continuous but not C^1, cf. [61], [62].

For a more detailed discussion of the counter-examples we refer to [22].

We can conclude that (except in dimension $n = 2$, as we shall see) vector valued minimizers or solutions of nonlinear elliptic systems are in general *nonregular* and we can only hope to prove *"partial regularity"*, i.e. regularity except possibly on a *closed "singular set"* (hopefully not too large).

Partial regularity results for solutions of nonlinear elliptic systems were proved during the years 1968–71 by Morrey [56], Giusti-Miranda [41], Giusti [37] and Pepe [64] on the basis of an indirect argument similar to the one introduced by De Giorgi and Almgren in the regularity theory of parametric minimal surfaces.

In 1978/79 the study of partial regularity has got a new impulse from the papers of Giaquinta-Giusti [25] and Giaquinta-Modica [28], [29], where a new argument of direct type (essentially a perturbation argument) was introduced on the basis of a higher integrability of the gradient (reverse Hölder-inequality with increasing support, see [28]).

It is not the aim of this chapter to give a report on these results; for that we refer to [22]. Here we shall confine ourselves to describe in the first part the techniques in a few very simple situations and in the second part to describe a few results in connection with the problem of the regularity of harmonic maps between Riemannian manifolds. For more information on this last topic we refer to Eells-Lemaire [14], [15] and Hildebrandt [43], [44].

6.1 A simple case

Let $u \in H^{1,2}(\Omega, \mathbb{R}^N)$ be a minimizer of the functional

$$\int_\Omega F(Du)dx$$

for which we assume:

(i) $\lambda|p|^2 \leq F(p) \leq \Lambda|p|^2$ for some positive constants λ and Λ.

(ii) $F \in C^2(\Omega)$ and $|F_{pp}| \leq L =$ constant
$F_{p_\alpha p_\beta}\xi_\alpha\xi_\beta \geq \nu|\xi|^2$ for $\forall\xi \in \mathbb{R}^N$, $\nu > 0$,

then we have

Theorem 6.1 *There exists an open subset* $\Omega_0 \subset \Omega$, *such that* $u \in C^{1,\alpha}(\Omega_0)$,

$$\Omega \backslash \Omega_0 = \left\{ x \in \Omega \mid \liminf_{R \to 0} \fint_{B_R(x)} |Du - (Du)_R|^2 \, dy > 0 \right\}$$

and $\dim_{\mathcal{H}}(\Omega \backslash \Omega_0) \leq n-2$.
($\dim_{\mathcal{H}}$ denotes the Hausdorff-dimension which will be explained in the sequel, $\Omega \setminus \Omega_0$ is called the *singular set*).

PROOF:

$$\begin{aligned} F(p) &= F(p_0) + F_{p^i_\alpha}(p_0)\left(p^i_\alpha - p^i_{0_\alpha}\right) \\ &\quad + \int_0^1 (1-t) F_{p^i_\alpha p^j_\beta}(tp + (1-t)p_0)\left(p^i_\alpha - p^i_{0_\alpha}\right)\left(p^j_\beta - p^j_{0_\beta}\right) dt \,, \end{aligned}$$

to get an approximation for this F, we consider

$$G(p) = F(p_0) + F_{p^i_\alpha}(p_0)\left(p^i_\alpha - p^i_{0_\alpha}\right) + \int_0^1 (1-t) F_{p^i_\alpha p^j_\beta}(p_0)\left(p^i_\alpha - p^i_{0_\alpha}\right)\left(p^j_\beta - p^j_{0_\beta}\right) dt \;.$$

Note that $|F(p) - G(p)| \leq \omega(|p_0|^2)|p - p_0|^2$!

Now we choose $p_0 := (Du)_{x_0,R}$ and look at the unique solution

$$v \in H^{1,2}(B_{R/2}(x_0)) \quad \text{of} \quad \begin{cases} \int_{B_{R/2}} G(Dv)dx \longrightarrow \text{minimize} \\ v - u \in H^{1,2}_0(B_{R/2}) \,. \end{cases}$$

As v is the solution of an elliptic second order system with constant coefficients, we get from L^p-theory (cf. 4.3, $f = Du - p_0$):

$$\int_{B_{R/2}} |Dv - p_0|^2 dx \leq c \int_{B_{R/2}} |Du - p_0|^2 dx \qquad 2 \leq s < \infty$$

and moreover

$$\int_{B_\rho} |Dv - (Dv)_\rho|^2 \, dx \leq c(\rho/R)^{n+2} \int_{B_{R/2}} |Dv - (Dv)_\rho|^2 \, dx \qquad \text{for all } \rho \leq R/2.$$

Thus ($p_0 := (Dv)_R$) we have

$$\int_{B_\rho} |Du - (Du)_\rho|^2\, dx \le c(\rho/R)^{n+2} \int_{B_{R/2}} \left|Du - (Du)_{R/2}\right|^2 dx + c \int_{B_{R/2}} |D(u-v)|^2 dx \ . \tag{6.1}$$

We have to estimate the second term:

$$\begin{aligned}
&\int_{B_{R/2}} |D(u-v)|^2 dx \\
&\le \frac{2}{\nu} \int_{B_{R/2}} [F(Dv) - F(Du)]dx \\
&\le \int_{B_{R/2}} [F(Dv) - G(Dv)] + \underbrace{[G(Dv) - G(Du)]}_{\le 0} + [G(Du) - F(Du)]dx \\
&\le c \int_{B_{R/2}} \omega\left(|Dv - p_0|^2\right)|Dv - p_0|^2 + \omega\left(|Du - p_0|^2\right)|Du - p_0|^2 dx \\
&\le c\left(\int_{B_{R/2}} |Dv - p_0|^{2^*} dx\right)^{2/2^*}\left(\int_{B_{R/2}} \omega\left(|Dv - p_0|^2\right)dx\right)^{2/n} \\
&\quad + c\left(\int_{B_{R/2}} |Du - p_0|^{2^*} dx\right)^{2/2^*}\left(\int_{B_{R/2}} \omega\left(|Du - p_0|^2\right)dx\right)^{2/n}
\end{aligned}$$

where we have used the fact that ω is bounded and $\omega^{n/2} = \omega \cdot \omega^{n/2-1}$. Now

$$\left(\int_{B_{R/2}} |Dv - p_0|^{2^*} dx\right)^{2/2^*} \underset{\text{Sobolev}}{\le} \int_{B_{R/2}} |D^2 u|^2 dx \underset{\text{Caccioppoli}}{\le} \frac{c}{R^2}\int_{B_R} |Du - p_0|^2 dx$$

so that we get in conclusion

$$\begin{aligned}
&\int_{B_{R/2}} |D(u-v)|^2 dx \\
&\le \frac{c}{R^2}\int_{B_R} |Du - p_0|^2 \left[\int_{B_{R/2}} \omega\left(|Du - p_0|^2\right) + \omega\left(|Dv - p_0|^2\right)dx\right]^{2/n} dx \qquad (6.2) \\
&\underset{\text{Jensen}}{\le} c\int_{B_R} |Du - p_0|^2 dx \left[\omega\left(\fint_{B_R} |Du - p_0|^2 dx\right)\right]^{2/n}
\end{aligned}$$

Putting (6.1) and (6.2) together we have

$$\begin{aligned}&\fint_{B_\rho} |Du-(Du)_\rho|^2\,dx\\ &\le c\left[(\rho/R)^2+(R/\rho)^n\omega\left(\fint_{B_R}|Du-(Du)_R|^2\,dx\right)\right]\fint_{B_R}|Du-(Du)_R|^2\,dx\ .\end{aligned}$$

Now we set $\Phi(x_0,\rho) := \fint_{B_\rho} |Du-(Du)_\rho|^2\,dx$.

Then

$$\Phi(x_0,\rho)\le c\left[(\rho/R)^2+(R/\rho)^n\omega\left(\Phi(x_0,R)\right)\right]\Phi(x_0,R)$$

and for any $\tau\in(0,1)$:

$$\Phi(x_0,\tau R)\le c\tau^2\left[1+\tau^{-n-2}\omega(\Phi(x_0,R))\right]\Phi(x_0,R)\ .$$

We split $\tau^2=\tau^{2\alpha}\tau^{2-2\alpha}$ for $\alpha\in(0,1)$ fixed, and choose τ in such a way that $2c\tau^{2-2\alpha}=1$.

Suppose that $\Phi(x_0,R)<\varepsilon$ so that

$$\tau^{-n-2}\omega\left(\Phi(x_0,R)\right)<1\ ,$$

then

$$\Phi(x_0,\tau^kR)\le\tau^{2\alpha k}\Phi(x_0,R)$$

thus $\Phi(x_0,\rho)\le c\rho^{2\alpha}$.

As $\Phi(x_0,R)$ is continuous in x_0, we conclude that the last inequality holds in a neighbourhood of x_0 and therefore that Du is Hölder-continuous in a neighbourhood of x_0, which proves the first part of the theorem.

Actually $D^2u\in L^2(\Omega)$ (see chapter 2) and hence

$$\fint_{B_R}|Du-(Du)_\rho|^2\le\frac{1}{R^{n-2}}\int_{B_R}|D^2u|^2dx\ ,$$

thus

$$\Omega\backslash\Omega_0\subset\left\{x\in\Omega\mid\liminf_{R\to0}\frac{1}{R^{n-2}}\int_{B_R(x)}|D^2u|^2dy>0\right\}\ .$$

The second part of the theorem follows from this remark and the next theorem (p. 106) which needs some preparation.

We introduce the notion of *Hausdorff-measure:*

Let X be a metric space and J a family of subsets of X such that $\emptyset\in J$.

Given a function $\xi : J \to [0,\infty]$, $\xi(\emptyset) = 0$ we set for all subsets $E \subset X$

$$\mu_\varepsilon(E) := \inf \left\{ \sum_{n=0}^{\infty} \xi(F_n) \mid F_n \in J\ ,\ \ E \subset \bigcup_{n=0}^{\infty} F_n\ ,\ \ \operatorname{diam} F_n < \varepsilon \right\}$$

and

$$\mu(E) := \lim_{\varepsilon \to 0} \mu_\varepsilon(E) \qquad (= \sup_{\varepsilon > 0} \mu_\varepsilon(E) \ \text{ as } \ \varepsilon < \delta \ \text{ implies } \ \mu_\varepsilon \geq \mu_\delta)$$

μ is an exterior measure and is called *the result of Caratheodori's construction for ξ and J. The k-dimensional Hausdorff-measure* corresponds to choosing $X = \mathbb{R}^n$

$J =$ the open sets in $\mathbb{R}^n$

$\xi(F) = \omega_k 2^{-k}\ (\operatorname{diam} F)^k$

where ω_k is the measure of the unit ball in $\mathbb{R}^n$.

$$\mathcal{H}^k(E) := 2^{-k}\omega_k \sup_{\varepsilon>0} \inf \left\{ \sum_{n=0}^{\infty} (\operatorname{diam} F_n)^k \right\}\ ,$$

$\{F_n\}$ is a countable family of open sets, $E \subset \bigcup\limits_{n=0}^{\infty} F_n$, $\operatorname{diam} F_n < \varepsilon$.

$\mathcal{H}^0(E) :=$ number of points of E.

It is not difficult toshow that, if $\mathcal{H}^k(E) < \infty$, then $\mathcal{H}^{k+\varepsilon}(E) = 0$ for all $\varepsilon > 0$.

If $k \in \mathbb{N}$, $\mathcal{H}^k$ is the Lebesgue-measure.

Finally we define the *Hausdorff-dimension of E:*

$$\dim_{\mathcal{H}} E := \inf \left\{ k \in \mathbb{R}^+ \mid \mathcal{H}^k(E) = 0 \right\} \ .$$

We have the

Theorem 6.2 *Let Ω be open in $\mathbb{R}^n$, $v \in L^1_{\text{loc}}(\Omega)$, $0 \leq \alpha < n$ and $E_\alpha :=$ $\left\{ x \in \Omega \mid \limsup_{\rho \to 0^+} \rho^{-\alpha} \int_{B_\rho(x)} |v| dy > 0 \right\}$.*

Then $\mathcal{H}^\alpha(E_\alpha) = 0$.

PROOF: It is sufficient to show that for all K compact and $K \subset \Omega$ holds $\mathcal{H}^\alpha(F) = 0$ where $F = E_\alpha \cap K$. Set

$$F^{(s)} := \left\{ x \in F \mid \limsup_{\rho \to 0} \rho^{-\alpha} \int_{B_\rho(x)} |v| dy > s^{-1} \right\}$$

then $F = \bigcap_{s=1}^{\infty} F^{(s)}$ and it is sufficient to show that $\mathcal{H}^\alpha(F^{(s)}) = 0$ for all s.

Let Q be an open subset of Ω with $K \subset Q \subset \bar{Q} \subset \Omega$ and $d := \operatorname{dist}(K, \partial Q)$. If we now fix $\varepsilon > 0$, then for all $x \in F^{(s)}$ there exists $0 < r(x) < \varepsilon < \delta$, such that

$$r(x)^{-\alpha} \int\limits_{B(x,r(x))} |v| dy > \frac{1}{2s} \quad .$$

From Besicovitch's theorem it follows that there exists a countable family $\{x_i\} \subset F^{(s)}$ with

$$B\left(x_i, r(x_i)\right) \cap B\left(x_j, r(x_j)\right) = \emptyset \qquad (i \neq j)$$

$\bigcap\limits_i B(x_i, 3r(x_i)) \supset F^{(s)}$.

Setting $r_i := r(x_i)$, we get

$$\sum_i r_i^\alpha \leq 2s \sum_i \int\limits_{B(x_i,r_i)} |v(y)| dy \leq 2s \int\limits_{\bigcup\limits_i B(x_i,r_i)} |v| dy \ .$$

Now since $\alpha < n$, we have

$$\operatorname{meas}\left\{\bigcup_i B(x_i, r_i)\right\} = \omega_n \sum_i r_i^n \leq \omega_n \varepsilon^{n-\alpha} \sum_i r_i^\alpha \leq 2s\omega_n \varepsilon^{n-\alpha} \int\limits_Q |v| dy \ ;$$

thus we conclude (ε is arbitrary small), that $\operatorname{meas}\left\{\bigcup\limits_i B(x_i, r_i)\right\}$ is small and therefore that $\sum\limits_i r_i^\alpha$ is arbitrary small, which implies $\mathcal{H}^\alpha(F^{(s)}) = 0$.

6.2 Reverse Hölder-inequality with increasing support

As we shall see in the sequel the following theorem play an important role in proving partial regularity.

Theorem 6.3 *Let Q be an n-dimensional cube and $f \in L^q_{\text{loc}}(Q)$ for $q > 1$, $f \geq 0$. Suppose that*

$$\left(\fint\limits_{Q_{R/2}(x_0)} f^q dx \right)^{1/q} \leq b \fint\limits_{Q_R(x_0)} f dx \qquad \textit{for all } s_0 \in Q \textit{ and } R < \operatorname{dist}(x_0, \partial Q)$$

(b a positive constant). Then $f \in L^p_{\text{loc}}(Q)$ for $p \in (q, q+\varepsilon)$ for some $\varepsilon > 0$. Moreover for all $Q_r \subset Q$ we have

$$\left(\fint_{Q_{r/2}} f^p dx \right)^{1/p} \leq c(b,q,n,\varepsilon) \left(\fint_{Q_r} f^q dx \right)^{1/q} .$$

This theorem, due to Giaquinta-Modica [28] extends an earlier result of Gehring [21] and can be considered as a local version of it. For extensions and comments we refer to [22].

PROOF: We start with the Whitney-decomposition:

$$\begin{aligned} Q_1 &:= \{x \in \mathbb{R}^n \mid |x_i| < 3/2, \ i = 1, \ldots, n\} \\ C_0 &:= \{x \in \mathbb{R}^n \mid |x_i| < 1/2, \ i = 1, \ldots, n\} \\ C_k &:= \left\{x \in Q_1 \mid 2^{-k} \leq \text{dist}(x, \partial Q_1) < 2^{-k+1}\right\} , \quad k \geq 1 \\ Q_1 &= C_0 \cup \left(\bigcup_{k \geq 1} C_k \right) . \end{aligned}$$

The maximal cubes contained in C_k are denoted by $P_{k,j}$ and if Q is a cube we denote by $\tilde{Q}$ the cube with the same center but double diameter. Finally, for $x \in Q_1$, we set

$$\Phi(x) := (\text{dist}(x, \partial Q_1))^n .$$

Note that on C_k and on $C_{k-1} \cup C_k \cup C_{k+1}$ we have $\Phi(x) \sim |P_{k,j}|$. Let $\lambda \geq \gamma_0 \int_{Q_1} f dx$, where γ_0 is to be chosen. Now for fixed k and j:

$$\begin{aligned} \lambda &\geq \gamma_0 \int_{\tilde{P}_{k,j}} f dx \\ &\geq \gamma_0 2^n |P_{k,j}| \fint_{\tilde{P}_{k,j}} f dx \\ &\geq \frac{\gamma_0 2^n}{b} \left(\fint_{P_{k,j}} f^q dx \right)^{1/q} |P_{k,j}| \\ &\geq \frac{\gamma_0 2^n}{b} c_1 \left(\fint_{P_{k,j}} (f\Phi)^q dx \right)^{1/q} \end{aligned}$$

i.e. $\lambda \geq \left(\fint_{P_{k,j}} (f\Phi)^q dx \right)^{1/q}$ if we choose $\gamma_0 \geq \frac{b}{2^n c_1}$.

Now we apply Calderon-Zygmund-argument to each $P_{k,j}$ and conclude, that there exists another sequence of cubes Q^j_k with disjoint interior, such that:

(i) $\lambda^q < \fint_{Q_k^j} (f\Phi)^q dx \le 2^n \lambda^q$

(ii) $f\Phi \le \lambda$ on $Q_1 \setminus \bigcup_{j,k} Q_k^j$

(iii) $Q_k^j \subset C_k$.

We have

$$\int_{\{x \in Q_1 | f\Phi > \lambda\}} (f\Phi)^q dx \le \sum_{j,k} |Q_j^j| \fint_{Q_k^j} (f\Phi)^q dx \overset{(i)}{\le} 2^n \lambda^q \sum_{j,k} |Q_k^j| \ .$$

The next step is to estimate $|Q_k^j|$. For that purpose let Q be *one* of the Q_k^j. From (i) we get

$$\begin{aligned} \lambda &\le \left(\fint_Q (f\Phi)^q dx \right)^{1/q} \\ &\le c|P_k^i| \left(\fint_Q f^q dx \right)^{1/q} \\ &\le c|P_k^j| \fint_Q f dx \\ &\le c \fint_{\tilde{Q}} (f\Phi) dx \ , \end{aligned}$$

i.e.

$$\lambda|\tilde{Q}| \le c \int_{\tilde{Q}} f\Phi dx \le c \int_{\tilde{Q} \cap \{f\Phi > \lambda\beta\}} f\Phi dx + c\beta\lambda|\tilde{Q}| \ ,$$

thus if β is small enough $(1 - c\beta > 0)$, we can write

$$\lambda|\tilde{Q}| \le \frac{c}{1 - c\beta} \int_{\tilde{Q} \cap \{f\Phi > \lambda\beta\}} f\Phi dx \ .$$

The $\tilde{Q}_k^j$ might intersect too many times so we have to use a covering argument (see [71, prop. 9]), which says, that from the covering $\{\tilde{Q}_k^j\}$ we can extract a sequence of *disjoint* cubes $\{F_k^j\}$, such that

$$\text{meas} \left(\bigcup_{j,k} \tilde{Q}_k^j \right) \le 5^n \sum_{j,k} |F_k^j| \ .$$

Therefore we conclude

$$(*)\qquad \int\limits_{\{x\in Q_1 | f\Phi>\lambda\}} (f\Phi)^q dx \leq c\lambda^{q-1} \int\limits_{\{x\in Q_1 | f\Phi>\beta\lambda\}} f\Phi dx$$

for all $\lambda \geq \gamma_0 \int\limits_{Q_1} f dx$.

From that as

$$\int\limits_{\{\beta\lambda<f\Phi\leq\lambda\}} (f\Phi)^q dx \leq c\lambda^{q-1} \int\limits_{\{f\Phi>\beta\lambda\}} f\Phi dx \ ,$$

we find

$$\int\limits_{\{x\in Q_1 | f\Phi>t\}} (f\Phi)^q dx \leq ct^{q-1} \int\limits_{\{x\in Q_1 | f\Phi>t\}} f\Phi dx \qquad \text{for all } t \geq \frac{\gamma_0}{\beta} \int\limits_{Q_1} f dx.$$

The result we claimed now follows from the following lemma due to Gehring [21]. We introduce

$$h(t) := \int\limits_{\{x\in Q_1 | f\Phi>t\}} f\Phi dx$$

and observe that

$$\int\limits_{\{f\Phi>t\}} (f\Phi)^p dx = -\int\limits_t^\infty s^{p-1} dh(s) \ .$$

Lemma 6.1 *Let $h(t)$ be a nonincreasing function*

$$h : [t_0, \infty) \to [0, \infty)$$

with $\lim\limits_{t\to\infty} h(t) = 0$.
Suppose moreover that we have for all $t \geq t_0$

$$-\int\limits_t^\infty s^{q-1} dh(s) \leq at^{q-1} h(t) \qquad (a = \text{ constant } > 1) \ ,$$

then

$$-\int\limits_{t0}^\infty s^{p-1} dh(s) \leq \frac{q-1}{a(q-1)-(a-1)(p-1)} t_0^{p-q} \left(-\int\limits_{t_0}^\infty s^{q-1} dh(s) \right)$$

for $p-1 \in \left[q-1, (q-1)\frac{a}{a-1}\right]$.

PROOF: It suffices to prove the theorem for $t_0 = 1$, then a rescaling argument gives the general result. We suppose that $h(s) = 0$ for all $s \in [k, \infty)$ (the general case can be handled by approximation, see [21], [22]).

For all $r > 0$ we set $I(r) := -\int_1^\infty s^r dh(s) = -\int_1^k s^r dh(s)$. Then

$$I(p-1) = -\int_1^k s^{p-q} s^{q-1} dh(s)$$

$$= I(q-1) + (p-q) \underbrace{\int_1^k s^{p-q-1} \left(-\int_s^k t^{q-1} dh(t) \right) dx}_{=:J} \quad ;$$

↑ integrating by parts

now by assumption

$$J \le a \int_1^k s^{p-2} h(s) ds$$

$$= -a\tfrac{1}{p-1} h(1) + \tfrac{a}{p-1} \int_1^k s^{p-1} dh(s)$$

$$\le \tfrac{a}{p-1} \int_1^k s^{p-1} dh(s) - \tfrac{1}{p-1} I(q-1) \ ,$$

so that we finally have

$$I(p-1) \le I(q-1) + a\frac{p-q}{p-1} I(p-1) - \frac{p-q}{p-1} I(q-1)$$

i.e.

$$\left[1 - a\frac{p-q}{p-1}\right] I(p-1) \le \frac{q-1}{p-1} I(q-1)$$

and the lemma is proved.

Given a functional $\mathbb{F}$ we call u a *spherical quasi-minimum* iff

$$\mathbb{F}(u; B_R) \le Q\mathbb{F}(u+\phi; B_R) \quad \text{for} \quad \forall \phi \in H_0^{1,2}(B_R) \quad \text{and} \quad \forall B_R \subset \Omega \ .$$

Note that any Q-minimum is a spherical Q-minimum; but in [30] it is shown that the converse is not true.

We have

Lemma 6.2 *Let u be a spherical quasi-minimum for the functional*

$$\mathbb{F}(u; \Omega) := \int_\Omega F(x, u, Du) dx$$

with $c_0|p|^2 \le F(x,u,p) \le c_1|p|^2$ *(c_0, c_1 positive constants), then* $u \in H^{1,p}_{\mathrm{loc}}(\Omega)$ *for some* $p > 2$ *and*

$$\left(\fint_{B_R} |Du|^p dx\right)^{1/p} \le \left(\fint_{B_{2R}} |Du|^2\right)^{1}/2 \quad \text{for } \forall B_{2R} \subset\subset \Omega\ .$$

PROOF: We choose as test function $\phi := u - \eta(u - u_R)$; then

$$\begin{aligned}
\int_{B_\rho} |Du|^2 dx &\le \int_{B_R} |Du|^2 dx \\
&\le Q \int_{B_R} |D(u - \eta(u - u_R))|^2\, dx \\
&\le c \int_{B_R} (1-\eta)^2 |Du|^2 dx + c \int_{B_R} |D\eta|^2 |u - u_R|^2 dx \\
&\le c \int_{B_R \setminus B_\rho} |Du|^2 dx + \frac{c}{(R-\rho)^2} \int_{B_R} |u - u_R|^2 dx
\end{aligned}$$

$\eta \equiv 1$ on B_ρ and $\eta \in C_0^\infty(B_R)$

which implies

$$\int_{B_\rho} |Du|^2 dx \le \frac{c}{c+1} \int_{B_R} |Du|^2 dx + \frac{c}{(R-\rho)^2} \int_{B_R} |u - u_R|^2 dx\ ,$$

so that (by lemma 5.1) we can omit the first term on the right and we get

$$\int_{B_R} |Du|^2 dx \le \frac{c}{R^2} \int_{B_{2R}} |u - u_R|^2 dx \le \frac{c}{R^2} \left(\int_{B_{2R}} |Du|^{2_*}\right)^{2/2_*}$$

(where $(2_*)^* = 2$, i.e. $2_* < 2$) or

$$\left(\fint_{B_R} |Du|^2 dx\right)^{1/2} \le c \left(\fint_{B_{2R}} |Du|^{2_*}\right)^{1/2_*} .$$

From theorem 6.3 there exists $p > 2$ such that

$$\left(\fint_{B_R} |Du|^p dx\right)^{1/p} \le c \left(\fint_{B_{2R}} |Du|^2 dx\right)^{1/2} .$$

Remark: Note that any weak solution of an elliptic quasilinear (or even nonlinear [30]) system is a spherical quasiminima. In fact if

$$\int_{\Omega} A(u)DuD\phi dx = 0 \qquad \text{for } \forall\phi \in H_0^{1,2}(\Omega)$$

taking $\phi = u - v$ with $\text{supp}(u - v) \subset\subset B_R$, wc gct

$$\begin{aligned} \nu \int_{B_R} |Du|^2 dx &\leq \int_{B_R} A(x,u)DuDudx \\ &= \int_{B_R} A(x,u)DuDvdx \\ &\leq c\left(\int_{B_R} |Du|^2 dx\right)^{1/2} \left(\int_{B_R} |Dv|^2 dx\right)^{1/2}, \end{aligned}$$

i.e.

$$\int_{B_R} |Du|^2 dx \leq c \int_{B_R} |Dv|^2 dx$$

thus L^p-estimates of the gradient hold for any minimizer or solutions of elliptic systems.

6.3 Some model results

Theorem 6.4 *Consider a weak solution of*

$$\int_{\Omega} A_{ij}^{\alpha\beta}(x,u)D_\alpha u^i D_\beta \phi^j dx = 0 \qquad \textit{for } \forall\phi \in H_0^{1,2}(\Omega, \mathbb{R}^N)$$

where

(i) $\left|A_{ij}^{\alpha\beta}(x,u)\right| \leq L =$ *constant*

(ii) $A_{ij}^{\alpha\beta}\xi_\alpha^i\xi_\beta^j \geq |\xi|^2$

(iii) the $A_{ij}^{\alpha\beta}$ *are (uniformly) continuous.*

Then there exists an open set Ω_0 *such that* $u \in C^{0,\alpha}(\Omega_0)$ *(if the* $A_{ij}^{\alpha\beta}$ *are Hölder-continuous, then* $u \in C^{1,\alpha}(\Omega_0)$*).*

Moreover the singular set $\Omega \setminus \Omega_0$ *is contained in*

$$\left\{ x \in \Omega \mid \liminf_{\rho \to 0} \fint_{B_\rho} |u - u_\rho|^2 dx > 0 \right\}$$

and thus $\dim_{\mathcal{H}} \Omega \setminus \Omega_0 < n - 2$.

PROOF: We fix $x_0 \in \Omega$. Then

$$(+) \qquad \int_{B_R(x_0)} A(x_0, u_R) Du D\phi dx = \int_{B_R(x_0)} [A(x_0, u_R) - A(x, u)] Du D\phi dx$$

for $\forall \phi \in H_0^{1,2}(B_R, \mathbb{R}^N)$.

We now split $u = v + (u - v)$ where v is the solution of

$$\int_{B_R(x_0} A(x_0, u_R) Dv D\phi dx = 0 \quad \text{and} \quad v = u \quad \text{on } \partial B_R(x_0) \ .$$

As usual

$$\int_{B_\rho} |Du|^2 dx \leq c(\rho/R)^n \int_{B_R} |Du|^2 dx + c \int_{B_R} |D(u - v)|^2 dx \ .$$

In (+) we can take $\phi = u - v$ and get

$$\begin{aligned} \int_{B_R} |D(u - v)|^2 dx &\leq \int_{B_R} [A(x_0, u_R) - A(x, u)]^2 |Du|^2 dx \\ &\leq \int_{B_R} \omega\left(|x - x_0|^2 + |u - u_R|^2\right) |Du|^2 dx \end{aligned}$$

and that is exactly the same estimate as in the linear case, but this time we do not know that ω is small (ω is the modulus of continuity i.e. bounded, concave and $\omega(t) \to 0$ for $t \to 0$). We thus continue the above estimate

$$\int_{B_R} \omega\left(|x - x_0|^2 + |u - u_R|^2\right) |Du|^2 dx$$

$$\leq c \left(\fint_{B_R} |Du|^p dx \right)^{2/p} \left(\fint_{B_R} \omega\left(|x - x_0|^2 + |u - u_R|^2\right) dx \right)^{1-2/p} |B_R|$$

Hölder

$$\leq \int_{B_{2R}} |Du|^2 dx \omega \left(R^2 + \fint_{B_R} |u - u_R|^2 dx \right) ,$$

Jensen

thus

$$\int_{B_{2R}} |Du|^2 dx \le c\left[(\rho/R)^n + \omega\left(R^2 + \frac{1}{R^{n-2}}\int_{B_{2R}} |Du|^2 dx\right)\right]\int_{B_R} |Du|^2 dx$$

and the result follows exactly as in section 6.1, where this time

$$\Phi(x_0, R) := \frac{1}{R^{n-2}}\int_{B_R} |Du|^2 dx \ .$$

To estimate the Hausdorff-dimension of the singular set we compute

$$\int_{B_R(x_0)} |Du|^2 dx \le \left(\int_{B_R(x_0)} |Du|^p dx\right)^{2/p} |B_R|^{1-2/p}$$

thus

$$\begin{aligned}\frac{1}{R^{n-2}}\int_{B_R(x_0)} |Du|^2 dx &\le \frac{1}{|B_R|^{\frac{n-2}{n}}}\left(\int_{B_R(x_0)} |Du|^p dx\right)^{2/p} |B_R|^{1-2/p} \\ &= \left(\frac{1}{|B_R|^{\left(\frac{n-2}{n}-1+\frac{2}{p}\right)\frac{p}{2}}}\int_{B_R(x_0)} |Du|^p dx\right)^{2/p} \\ &\le c\left(\frac{1}{R^{n-p}}\int_{B_R(x_0)} |Du|^p dx\right)^{2/p}\end{aligned}$$

and we conclude

$$\begin{aligned}&\left\{x_0 \in \Omega \mid \liminf_{R\to 0} \frac{1}{R^{n-2}}\int_{B_R(x_0)} |Du|^2 dx > 0\right\} \\ &\subset \left\{x_0 \in \Omega \mid \liminf_{R\to 0} \tfrac{1}{R^{n-p}}\int_{B_R(x_0)} |Du|^p dx > 0\right\} =: E\end{aligned}$$

and $\mathcal{H}^{n-p}(E) = 0$ for some $p > 2$.

As a further example we consider a minimizer u of

$$(++) \qquad \int_\Omega A_{ij}^{\alpha\beta}(x,u) D_\alpha u^i D_\beta u^j dx \ ,$$

where the coefficients satisfy the same assumptions as before (cf. p. 113). Note that the functional is not differentiable. We here have the

Theorem 6.5 *Under the above assumptions any minimizer u of $(++)$ is Hölder-continuous (and if the $A_{ij}^{\alpha\beta}$ are Hölder-continuous, has also Hölder-continuous derivatives) in some open set Ω_0; Moreover* $\dim_{\mathcal{H}}(\Omega \setminus \Omega_0) < n - 2$.

PROOF: Let $v \in H_0^1(B_R, \mathbb{R}^N)$ be a minimizer of

$$\int_{B_R} A(x_0, u_{x_0,R}) D_\alpha v^i D_\beta v^j dx \quad \text{and} \quad v - u \in H_0^1(B_R, \mathbb{R}^N)$$

(where u minimizes $(++)$). Because of L^p-theory

$$\int_{B_R} |Dv|^p dx \leq c \int_{B_R} |Du|^p dx$$

and

$$\int_{B_\rho} |Du|^2 dx \leq c(\rho/R)^n \int_{B_R} |Du|^2 dx + c \int_{B_R} |D(u-v)|^2 dx \ .$$

Then

$$\begin{aligned} \int_{B_R} |D(u-v)|^2 dx \ \leq \ & \int_{B_R} \omega\left(|x-x_0|^2 + |u-u_R|^2\right) |Du|^2 dx \\ & + \int_{B_R} \omega\left(|x-x_0|^2 + |v-u_R|^2\right) |Dv|^2 dx \end{aligned}$$

and the proof continues as the proof of theorem 6.1.

Finally we consider a general functional

$$\int_\Omega F(x, u, Du) dx \ .$$

Suppose F to be of class C^2 with respect to p and

(i) $|F_{pp}| \leq L =$ constant

(ii) $F_{p_\alpha p_\beta} \xi^\alpha \xi^\beta \geq \nu |\xi|^2$ for $\forall \xi \in \mathbb{R}^N$, $\nu > 0$.

(iii) $|p|^{-2} F(x, u, p)$ is Hölder-continuous with respect to (x, u) uniformly in p,

then we have the following theorem which we state without proof (see [31] and [23]):

Theorem 6.6 *Let u be a minimizer of the above functional, then there exists an open set $\Omega_0 \subset \Omega$, such that $u \in C^{1,\alpha}(\Omega_0)$ and moreover meas$(\Omega \setminus \Omega_0) = 0$.*

Harmonic mappings for dimension $n = 1$ are geodesics, for dimension $n = 2$ minimal surfaces. Let us recall *some known results:*

In 1964 Eells-Sampson [16] proved that there *exist* harmonic maps between two Riemannian manifolds (M'^n, g) and (M^N, h), provided that the sectional curvature of M^N is nonnegative. Actually they proved that the associated parabolic flow has good properties and converges for the time going to ∞.

In 1977 Hildebrandt-Kaul-Widman [45] proved that minimizing maps in $H^{1,2}$ even in case of positive sectional curvature are regular, provided the images of the mappings one considers are contained in a geodesic ball $B_R(q)$, which does not intersect the cut-locus of q and has radius $R < \frac{\pi}{2\sqrt{k}}$ ($0 < k$ is an upper bound for the sectional curvature of M^N). They also proved existence of a regular energy minimizing map, if the boundary of M^n is mapped into $B_R(q)$.

Moreover they showed that the map $u^*(x) := \left(\frac{x}{|x|}, 0\right)$ from the unit ball of $\mathbb{R}^n$ into the equator of the standard-sphere in $\mathbb{R}^{n+1}$ is a critical point for the energy-functional $\mathbb{E}$.

In 1982 Jaeger-Kaul [46] showed that for $n \geq 7$ u^* is an absolute minimum for the energy, but that for $n \leq 6$ u^* is even unstable (i.e. the second variation is negative).

Thus energy minimizing maps can be *singular*. From the result we quoted earlier (cf. p. 116), we know that the maps $u : (\mathbb{R}^n, (G_{\alpha\beta}))) \to (\mathbb{R}^N, (g_{ij}))$ are regular in some open set Ω_0 and that $\dim_{\mathcal{H}}(\Omega \setminus \Omega_0) < n - 2$.

That this also holds in case the target manifold is not covered by just one chart has been proved by Schoen-Uhlenbeck [66].

In 1983 Baldes [2] proved the following:

The map

$$u^* : B_1(0) \to \left\{(y.y_{n+1}) \in \mathbb{R}^{n+1} \mid |y|^2 + \frac{y_{n+1}}{a^2} = 1\right\}$$

is stable if $a^2 > \frac{4(n-1)}{(n-2)^2}$ *and unstable if* $a^2 < \frac{4(n-1)}{(n-2)^2}$.

So in case $n = 3$ there are stable maps for the energy functional which are singular.

For the following theorem compare Giaquinta-Giusti [27] and for general maps Schoen-Uhlenbeck [66].

Theorem 6.7 *Let u be a bounded local minimizer of*

$$(\sim) \qquad \int_\Omega G^{\alpha\beta}(x) g_{ij}(u) D_\alpha u^i D_\beta u^j \, dx$$

with $(G^{\alpha\beta})$, (g_{ij}) symmetric positive definite matrices with smooth coefficients.

If $n = 3$, u has at most isolated singular points and in general the singular set has Hausdorff-dimension $d \leq n - 3$.

The proof relies on the following two facts:

Fact 1 (this actually is true for general functionals of the form

$$\int_\Omega A_{ij}^{\alpha\beta}(x,u)D_\alpha u^i D_\beta u^j dx\ ):$$

Let $A^{(\nu)}(x,z)$ be a sequence of continuous functions in $B_1(0)\times\mathbb{R}^N$ converging uniformly to $A(x,z)$ and

(i) $|A^{(\nu)}(x,z)| \le M =$ *constant*

(ii) $A^{(\nu)}_{\alpha\beta}\xi_\alpha\xi_\beta \ge |\xi|^2$ *for all* $\xi\in\mathbb{R}^n$

(iii) $|A^{(\nu)}(x,z) - A^{(\nu)}(x',z')| \le \omega\left(|x-x'|^2 + |z-z'|^2\right)$

ω bounded, continuous, concave and $\omega(0)=0$.

Now let $u^{(\nu)}$ be a local minimum of

$$\int_\Omega A^{(\nu)}(x,u^{(\nu)})Du^{(\nu)}Du^{(\nu)}dx$$

and assume that $u^{(\nu)} \longrightarrow v$ in $L^2(B_1)$. Then v is a local minimizer for

$$\int_\Omega A(x,v)DvDvdx\ .$$

Moreover, if x_ν is a singular point for $u^{(\nu)}$ and $x_\nu \to x_0$, then x_0 is a singular point for v.

The main ingredients for the proof of fact 1, see [27], are Caccioppoli-inequality for $u^{(\nu)}$ and the L^p-estimates for $Du^{(\nu)}$.

Fact 2 *(monotonicity-theorem)*

After a possible change of coordinates we have $0\in\Omega$ and $G^{\alpha\beta}(0)=\delta^{\alpha\beta}$. Moreover we assume that

$$(+)\qquad \int_0^1 \frac{\omega(t^2)}{t}dt < \infty\ .$$

Let v be a minimizer in $B_1(0)$. Then for $\rho<R<1$ we have

$$\int_{\partial B_1} |u(Rx)-u(\rho x)|^2 d\mathcal{H}^{n-1} \le \gamma\log(\rho/R)[\Phi(R)-\Phi(\rho)]$$

where $\Phi(t) = t^{2-n}\exp\left(c\int_0^t \frac{\omega(s^2)}{s}dx\right)\int_{B_t(0)} A(x,u)DuDudx.$

PROOF OF FACT 2: Suppose (for the sake of simplicity) that the coefficients do not depend on x. In this case

$$\Phi(t) = t^{2-n} \int_{B_t} A(u) DuDudx \ .$$

For $t < 1$ set $x_t := t\frac{x}{|x|}$, $u_t := u(x_t)$.

Then

$$\begin{aligned} &\mathcal{F}(u; B_t) \leq \mathcal{F}(u_t; B_t) \\ &= \int_{B_t} A(u(x_t)) \frac{t^2}{|x|^2} \left(\delta_{\alpha h} - \frac{x_\alpha x_h}{|x|^2}\right) \left(\delta_{\beta k} - \frac{x_\beta x_k}{|x|^2}\right) D_h u^i(x_t) D_k u^j(x_t) dx \\ &=: (*) \ . \end{aligned}$$

Note that

$$\int_{B_t} |x|^{-2} f(x)_t dx = \frac{t^{-1}}{n-2} \int_{\partial B_t} \int_{\partial B_t} f(x) d\mathcal{H}^{n-1}$$

thus

$$\begin{aligned} (*) &= \frac{t}{n-2} \left\{ \int_{\partial B_t} A(u) DuDud\mathcal{H}^{n-1} \right. \\ &\quad \left. - \int_{\partial B_t} A_{ij}^{\alpha\beta} \frac{x_\alpha x_h}{|x|^2} \left[2\delta_{\beta k} - \frac{x_\beta x_k}{|x|^2} \right] D_h u^i D_k u^j d\mathcal{H}^{n-1} \right\} \\ &= \frac{t}{n-2} \left\{ \int_{\partial B_t} A(u) DuDud\mathcal{H}^{n-1} - \int_{\partial B_t} \frac{|\langle x, Du \rangle|^2}{|x|^2} d\mathcal{H}^{n-1} \right\} \end{aligned}$$

where $\langle x, Du \rangle := x_\alpha D_\alpha u$.

Now $t^{2-n} \int_{\partial B_t} A(u) DuDud\mathcal{H}^{n-1} = \Phi'(t) + (n-2)\frac{\Phi(t)}{t}$ therefore

$$\Phi'(t) \geq t^{2-n} \int_{\partial B_t} \frac{|\langle x, Du \rangle|^2}{|x|^2} d\mathcal{H}^{n-1}$$

and integrating we get

$$\Phi(R) - \Phi(\rho) \geq \int_\rho^R t^{2-n} \int_{\partial B_t} \frac{|\langle x, Du \rangle|^2}{|x|^2} d\mathcal{H}^{n-1} dt \ .$$

On the other hand

$$\begin{aligned} |u(Rx)-u(\rho x)|^2 &\leq \left(\int_\rho^R |\langle x, Du(tx)\rangle| dt\right)^2 \\ &= \left(\int_\rho^R \sqrt{t}|\langle x, Du(tx)\rangle| \frac{1}{\sqrt{t}} dt\right)^2 \\ &\leq \log(R/\rho)\int_\rho^R t|\langle x, Du\rangle|^2 dt \ . \end{aligned}$$

Integrating over ∂B_1 we get the result.

PROOF OF THE THEOREM: Suppose that the theorem is not true and choose a sequence of singular points $\{x_\nu\}$ converging to $x_0 = 0$. We set $R_\nu := 1/2|x_\nu| < 1$, then $u^{(\nu)} := u(R_\nu x)$ is a local minimizer for

$$\mathbb{F}^{(\nu)}(u^{(\nu)}; B_1) := \int_{B_1} A^{(\nu)}(x, u^{(\nu)}) Du^{(\nu)} Du^{(\nu)} dx$$

where $A^{(\nu)}(x,z) = A(R_\nu x, z)$. Moreover each $u^{(\nu)}$ has a singular point y_ν with $|y_\nu| = 1/2$.

Since the $u^{(\nu)}$ are uniformly bounded we can suppose that they converge weakly in $L^2(B_1)$ to some v and also $y_\nu \to y_0$.

From fact 1 we conclude that v is a minimizer for $\int A(0,v)DvDvdy$ and that v is singular at $y_0(|y_0| = 1/2)$ and at 0.

Now $\Phi(t)$ defined in fact 2 is bounded, increasing and thus tends to a limit for $t \to 0$. By the Caccioppoli-inequality this limit is finite.

Moreover for $\rho = \lambda R_\nu$ and $R = \mu R_\nu, 0 < \lambda < \mu < 1$, we have

$$\int_{\partial B_1} \left|u^{(\nu)}(\lambda x) - u^{(\nu)}(\mu x)\right| d\mathcal{H}^{n-1} \leq c\log(\mu/\lambda) \underbrace{[\Phi(\mu R_\nu) - \Phi(\lambda R_\nu)]}_{\to 0(\nu\to\infty)}$$

Therefore we conclude

$$\int_{\partial B_1} |v(\lambda x) - v(\mu x)|^2 dx = 0 \qquad \text{for a.e. } \lambda \text{ and a.e. } \mu$$

i.e. v is homogeneous of degree zero.

This means that the whole segment joining 0 to y_0 consists of singular point, so that the singular set has Hausdorff-dimension at least 1. This is a contradiction because $n = 3$ implies $\dim_{\mathcal{H}}(\Omega \backslash \Omega_0) < 3 - 2 = 1$.

The second part of the theorem follows from an argument similar to the one used by Federer in [17], see [27].

Remarks:

1) The special structure of the coefficients (i.e. the splitting of variables) has only been used in fact 2!

2) Note that we have also proved:

   If minimizers of ($\sim$) are regular in dimension n, then they have at most isolated singularities in dimension $n+1$.

3) Energy minimizing maps with smooth Dirichlet boundary values have no singularity at the boundary, see [72], [67].

But we emphasize that in general solutions of quasilinear elliptic systems with smooth boundary values are singular at the boundary, cf. [73].

For $n \geq 7$ minimizers of $\mathbb{E}$ have singularities. We are now interested in the singularities of such map

$$U : \Omega \subset \mathbb{R}^n \longrightarrow S^{N-1} \subset \mathbb{R}^N$$

with image in the upper hemisphere.

If we introduce the coordinates induced by stereographic projection on S^{N-1} and u is a representation of U, the energy of u is given by

$$\mathbb{E}(u;\Omega) := \int_\Omega \frac{|Dz|^2}{(1+|u|^2)^2} dx \ .$$

Remarks:

1) For any bounded $v \in H^{1,2}(\Omega, \mathbb{R}^N)$ (i.e. v avoids a neighbourhood of the southpole) one can construct a new map $\tilde{v}$ with image of the upper half-sphere ("reflected map through the equator")

$$\tilde{v} := \begin{cases} v & \text{if} \quad |v| \leq 1 \\ \frac{v}{|v|^2} & \text{if} \quad |v| > 1 \end{cases}$$

   and $\mathbb{E}(v;\Omega) = \mathbb{E}(\tilde{v};\Omega)$.

2) If $u \in H^{1,2}(\Omega, \mathbb{R}^N)$, $|u| \leq 1$, minimizes the energy among all maps v with $v = 0$ on $\partial\Omega$ and image of v in the upper hemisphere (i.e. $|v| \leq 1$), then u minimizes the energy in the class

$$\{v \in H^{1,2}(\Omega, \mathbb{R}^N) \mid v \text{ bounded}, \quad v = u \text{ on } \partial\Omega\} \ .$$

Thus u is a solution of the Euler-equation

$$\int_{\Omega} \left[\frac{Du\ D\phi}{(1+|u|^2)^2} - 2\frac{|Du|^2}{(1+|u|^2)^2}\frac{u\cdot\phi}{1+|u|^2} \right] dx = 0$$

for $\forall \phi \in H_0^{1,2}(\Omega, \mathbb{R}^N) \cap L^\infty(\Omega, \mathbb{R}^N)$.

Out of u we construct (by the procedure in the proof of the theorem before) $\tilde{u}$, which is

1) energy-minimizing

2) homogeneous of degree zero

3) singular at $x_0 = 0$.

Then we choose the test function $\phi(x) = \tilde{u}(x)\eta(x)$ in the Euler-equation to get

$$\int_{\partial B_1} \frac{|D\tilde{u}|^2}{(1+|\tilde{u}|^2)^2} \cdot \frac{1-|\tilde{u}|^2}{1+|\tilde{u}|^2} d\mathcal{H}^{n-1} = 0 \ ;$$

thus either $|\tilde{u}| = 1$ or $\tilde{u}$ is constant and has no singularities.

In the first case we have produced an energy minimizing map homogeneous of degree one with $|\tilde{u}| = 1$ and singularity at zero. We now exclude this possibly if the dimension is $n \leq 6$:

We set $\mathbb{E}(t) := \mathbb{E}(u + t\phi; B_1)$; then

$$\delta E = \frac{d}{dt}\mathbb{E}(t)\Big|_{t=0} = 1/2 \int_{B_1} \left[DuD\phi - u\cdot\phi|Du|^2 \right] dx$$

$$\begin{aligned} \delta^2 E &= \frac{d^2}{dt^2}\mathbb{E}(t)\Big|_{t=0} \\ &= 1/2 \int_{B_1} \left[|D\phi|^2 - 4u\cdot\phi DuD\phi - |Du|^2|\phi|^2 + 3(u\cdot\phi)^2|Du|^2 \right] dx \ . \end{aligned}$$

In the second variation we insert $\phi(x) := u(x)|Du(x)|\eta(|x|)$ and find

$$\delta^2\mathbb{E} = 1/2 \int_{B_1} \left[c^2|D\eta|^2 - \eta^2 \left[c^4 + 1/2\Delta c^2 - |Dc|^2 \right] \right] dx \geq 0$$

where $c(x) := |Du(x)|$.

We have the following

main lemma (without proof, see [34])

$$c^4 + 1/2\Delta c^2 - |Dc|^2 \geq \frac{c^2}{|x|^2} + \frac{c^4}{n-1} \ .$$

From that we find

$$\int_{B_1} \left[c^2|D\eta|^2 - \eta^2 \left[\frac{c^2}{|x|^2} + \frac{c^4}{n-1} \right] \right] dx \geq 0$$

and we can say that either $c^2 \equiv 0$ (but then $u =$ constant and has no singularities) or

$$\int_0^1 r^{n-3}\dot{\eta} dr - \underbrace{\left(1 + \frac{1}{n-1} \frac{\int_{B_1} c^4 dH^{n-1}}{\int_{\partial B_1} c^2 dH^{n-2}} \right)}_{=:1+\sigma, \sigma>0} \int_0^1 r^{n-5}\eta^2 dr \geq 0$$

This implies

$$\int_0^\infty r^{n-3}\dot{\eta} dr - (1+\sigma) \int_0^\infty r^{n-5}\eta^2 dr \geq 0$$

for all η for which these integrals converge.

Hence, if we choose

$$\eta(r) := \begin{cases} r^\alpha & \text{for} \quad r < 1 \\ r^\beta & \text{for} \quad r \geq 1 \end{cases}$$

with $\alpha := 1/2(4-n+\varepsilon)$ and $\beta := 1/2(4-n-\varepsilon)$ we find $1/2(4-n)^2 \geq 2(1+\sigma)$ i.e. $n \geq 4 + 2\sqrt{1+\sigma}$.

Bibliography

[1] Agmon, S. – *Lectures on elliptic boundary value problems,* N.J. Van Nostrand, Princeton 1965.

[2] Baldes, A. – *Stability properties of the equator map from a ball into an ellipsoid,* Math. Z. **63** (1977), 337–403.

[3] Ball, J.M. – *Convexity conditions and existence theorems in nonlinear elasticity,* Arch. Rat, Mech. Anal. **63** (1977), 337–403.

[4] Ball, J.M., Curries, J.C., Olver, P.J. – *Null Lagrangians, weak continuity, and variational problems of arbitrary order,* J. Funct. Anal. **41** (1981), 135–174.

[5] Bers, L., Schechter, M. – *Elliptic equations, in: "Partial Differential Equations", 131–299,* Interscience, New York, 1964.

[6] Calderon, A.P., Zygmund, A. – *On the existence of certain singular integrals,* Acta Math. **88** (1982), 85–130.

[7] Campanato, S. – *Proprietà di Hölderianità di alcune classi die funzioni,* Ann. Sc. Norm. Sup. Pisa **17** (1963), 175–188.

[8] Campanato, S. – *Equazioni ellittiche del secondo ordine e spazi* $\mathcal{L}^{2,\lambda}$, Ann. Mat. Pura e Appl. **69** (1965), 321–380.

[9] Campanato, S. – *Su un teorema di interpolazione di G. Stampacchia,* Ann. Sc. Norm. Sup., Pisa **20** (1966), 649–652.

[10] Campanato, S., Stampacchia, G. – *Sulle maggiorazioni in* L^p *nella teoria delle equazioni ellittiche,* Boll. UMI **20** (1965), 393–399.

[11] De Giorgi, E. – *Sulla differenziabilità e l'analiticità delle estremali degli integrali multipli regolari,* Mem. Accad. Sci . Torino cl. Sci. Fis. Mat. Nat. (3) **3** (1957), 25–43.

[12] De Giorgi, E. – *Un esempio di estremali discontinue per un problema variazionale di tipo ellittico,* Boll. UMI **4** (1986), 135–137.

[13] DI BENEDETTO, E., TRUDINGER, N.S. – *Harnack inequality for quasiminima of variational integrals,* Annales de l'Institut H. Poincaré: Analyse Nonlinéaire **1** (1984), 295–308.

[14] EELLS, J., LEMAIRE, L. –, *A report on harmonic maps,* Bull. London Math. Soc. **10** (1978), 1–68.

[15] EELLS, J., LEMAIRE, L. –, *Selected topics in harmonic maps,* Conf. Board Math. Sci., Regional conference series in Math. 50, Am. Math. Soc., Rhode Island, 1983.

[16] EELLS, J., SAMPSON, J.H. – *Harmonic mappings of Riemannian manifolds,* Amer. J. Math. **86** (1964), 109–160.

[17] FEDERER, H. – *The singular set of area minimizing rectifiable currents with codimension one and area minimizing flat chains modulo two with arbitrary codimension,* Bull. Amer. Math. Soc. **76** (1970), 767–771.

[18] FREHSE, J. – *A discontinuous solution of a mildly nonlinear elliptic system,* Math. Z. 134 (1973), 229–230.

[19] FREHSE, J. – *A note on the Hölder continuity of solutions of variational problems,* Abhandlung Math. Sem. Hamburg **43** (1975), 59–63.

[20] GARDING, L. – *Dirichlet problem for linear elliptic partial differential equations,* Math. Scand. **1** (1953), 55–72.

[21] GEHRING, F.W. – *The L^p-integrability of the partial derivatives of a quasi conformal mapping,* Acta Math. **130** (1973), 265–277.

[22] GIAQUINTA, M. – *Multiple Integrals in the Calculus of Variations and Nonlinear Elliptic Systems,* Annals of Math. Studies No. 105, Princeton Univ. Press, Princeton 1983.

[23] GIAQUINTA, M., IVERT, P.-A. – *Partial regularity for minima of variational integrals,* Ank. for Math. **23** (1987), 221–229.

[24] GIAQUINTA, M. – *The regularity problem of extremals of variational integrals,* Proc. NATO/LMS Advanc. Study Inst. on "Systems of nonlinear partial differential equations" Oxford, July 25 – August 7, 1982, in: "Systems of nonlinear PDE", D. Reidel Publ. Co., 1983.

[25] GIAQUINTA, M., GIUSTI, E. – *Partial regularity for the solution to nonlinear parabolic systems,* Ann. Mat. Pura e Appl. **47** (1973), 253–266.

[26] GIAQUINTA, M., GIUSTI, E. – *On the regularity of the minima of variational integrals,* Acta Math. **148** (1982), 31–46.

[27] GIAQUINTA, M., GIUSTI, E. – *The singular set of the minima of certain quadratic functionals,* preprint 453 S.F.B. 72 Bonn (1981), Ann. Sc. Norm. Sup. Pisa **9** (1984).

[28] GIAQUINTA, M., MODICA, G. – *Regularity results for some classes of higher order nonlinear elliptic systems,* J. für reine und angew. Math. **311/312** (1979), 145–169.

[29] GIAQUINTA, M., MODICA, G. – *Almost-everywhere regularity results for solutions of nonlinear elliptic systems,* manuscripta math. **28** (1979), 109–158.

[30] GIAQUINTA, M., GIUSTI, E. – *Quasi-minima,* Annals de l'Institut H. Poincaré: Analyse Nonlinéaire, **1** (1984), 79–104.

[31] GIAQUINTA, M., GIUSTI, E. – *Differentiability of minima of nondifferentiable functionals,* Inventiones Math. **72** (1983), 285–298.

[32] GIAQUINTA, M., GIUSTI, E. – *Global $C^{1,\alpha}$-regularity for second-order quasilinear elliptic equations in divergence form,* J. für reine und angew. Math. **351** (1984), 55–65.

[33] GIAQUINTA, M., GIUSTI, E. – *Sharp estimates for the derivatives of local minima of variational integrals,* Boll. UMI (6) **3-A** (1984), 239–248.

[34] GIAQUINTA, M., SOUČEK, J. – *Harmonic maps into a hemisphere,* Ann. Sc. Norm. Sup. Pisa **12** (1985), 81–90.

[35] GILBARG, D., TRUDINGER, N.S. – *Elliptic partial differential equations of second order,* Springer Verlag, Heidelberg, New York (1977).

[36] GIUSTI, E. – *Precisazione delle funzioni $H^{1,p}$ e singolarità delle soluzioni deboli di sistemi ellittici non lineari,* Boll. UMI **2** (1969), 71–76.

[37] GIUSTI, E. – *Regolarità parziale delle soluzioni di sistemi ellittici quasi lineari di ordine arbitrario,* Ann. Sc. Norm. Sup. Pisa **23** (1969), 115–141.

[38] GIUSTI, E. – *Equazioni ellittiche del secondo ordine,* Quaderni dell'Un. Mat. Italiana **6** (1968) Ed. Pitagora, Bologna.

[39] GIUSTI, E. – *Regularity and singularities of solutions of nonlinear elliptic systems,* Proc. NATO/LMS Advanc. Study Inst. on "Systems of nonlinear partial differential equations" Oxford, July 25–August 7, 1982, In: "Systems of Nonlinear PDE", D. Reidel Publ. Co., 1983.

[40] GIUSTI, E., MIRANDA, M. – *Un esempio di soluzioni discontinue per un problema di minimo relativo ad un integrale regolare del calcolo delle variazioni,* Boll. UMI **2** (1968), 1–8.

[41] GIUSTI, E., MIRANDA, M. – *Sulla regolarità delle soluzioni deboli di una classe di sistemi ellittici quasilineari,* Arch. Rat. Mech. Anal. **31** (1986), 173–184.

[42] HILDEBRANDT, S. – *Liouville theorems for harmonic mappings and an approach to Bernstein theorem,* Annals of Math. Studies 102, Princeton (1982), 107–132.

[43] HILDEBRANDT, S. – *Nonlinear elliptic systems and harmonic mappings,* Vorlesungsreihe SFB 72 No. 3 (1980); Proc. Beijing Symp. Diff. Geo. and Diff. Eq., Gordon & Breach, New York, 1982.

[44] HILDEBRANDT, S. – *Elliptic systems of P.D.E.,* Proc. NATO/LMS Advanc. Study Inst. on "Systems of nonlinear partial differential equations" Oxford, July 25–August 7, In: "Systems of Nonlinear PDE", D. Reidel Publ. Co., 1983.

[45] HILDEBRANDT, S., KAUL, H., WIDMAN, K.-O. – *An existence theorem for harmonic mappings of Riemannian manifolds,* Acta Math. **138** (1977), 1–16.

[46] JÄGER, W., KAUL, H. – *Rotationally symmetric harmonic maps from a ball into a sphere and the regularity problem for weak solutions of elliptic systems,* J. Reine Angew. Math. **343** (1983), 146–161.

[47] JOHN, F., NIRENBERG, L. – *On functions of bounded mean oscillation,* Comm. Pure Appl. Math. **14** (1961), 415–426.

[48] KRYLOV, N.V., SAFANOV, M.V. – *Certain properties of solutions of parabolic equations with measurable coefficients,* Izvestia Akad. Nauk SSSR **40** (1980), 161–175, Engl. translation Math. USSR Izv. **16** (1981).

[49] LADYZHENSKAYA, O.A., URAL'TSEVA, N.N. – *Linear and quasilinear elliptic equations,* Moscow, Nauka (1964); Engl. Transl. Academic Press New York (1986), Second Russian edition: Nauka (1973).

[50] MAZ'YA, V.G. – *Examples of nonregular solutions of quasilinear elliptic equations with analytic coefficients,* Funktsional'nyi Analiz. i Ego Prilosheniya **2** (1968), 53–57.

[51] MEYERS, N.G. – *Quasi-convexity and lower semicontinuity of multiple variational integrals of any order,* Trans. Amer. Math. Soc. **119** (1965), 125–149.

[52] MORREY, C.B. JR. – *Multiple integral problems in the calculus of variations and related topics,* Univ. California Publ. Math. **1** (1943), 1–130.

[53] MORREY, C.B. JR. – *Quasi-convexity and the lower semicontinuity of multiple integrals,* Pacific J. Math. **2** (1952), 25–53.

[54] MORREY, C.B. JR. – *Second order elliptic systems of differential equations,* Ann. of Math. Studies No. 33, Princeton Univ. Press (1954), 101–159.

[55] MORREY, C.B. JR. – *Multiple integrals in the calculus of variations,* Springer Verlag, Heidelberg, New Yor, (1966).

[56] MORREY, C.B. JR. – *Partial regularity results for nonlinear elliptic systems,* Journ. Math. and Mech. **17** (1969), 649–670.

[57] MOSER J. – *A new proof of de Giorgi's theorem concerning the regularity problem for elliptic differential equations,* Comm. Pure Appl. Math. **13** (1960), 457–468.

[58] MOSER J. – *On Harnack's theorem for elliptic differential equations,* Comm. Pure Appl. Math. **14** (1961), 577–591.

[59] NASH, J. – *Continuity of solutions of parabolic and elliptic equations,* Amer. J. Math. **8** (1958), 931–954.

[60] NEČAS, J. – *Let Méthodes directes en théorie des équations elliptiques,* Praha, Akademia (1967).

[61] NEČAS, J. – *Example of an irregular solution to a nonlinear elliptic system with analytic coefficients and conditions for regularity, in: Theory of Non Linear Operators,* Abhandlungen Akad. der Wissen. der DDR (1977), Proc. of a Summer School held in Berlin (1975).

[62] NEČAS, J. – *On the regularity of weak solutions to variational equations and inequalities for nonlinear second order elliptic systems,* Equadiff IV, Praha, Springer Verlag Lecture Notes 702.

[63] NIRENBERG, L. – *Remarks on strongly elliptic partial differential equations,* Comm. Pure Appl. Math. **8** (1955), 649–675.

[64] PEPE, L. – *Risultati di regolarità parziale per le soluzioni $H^{1,p}(\Omega)$, $1 < p < 2$, di sistemi ellittici quasilineari,* Ann. Univ. Ferrara **8** (1971), 129–148.

[65] ROYDEN, H.L. – *Real Analysis,* Mac Millan, Toronto, 1968.

[66] SCHOEN, R., UHLENBECK, K. – *A regularity theory for harmonic maps,* J. Diff. Geo. **17** (1982), 307–335.

[67] SCHOEN, R., UHLENBECK, K. – *Boundary regularity theory and miscellaneous results on harmonic maps,* J. Diff. Geo. **17** (1982), 307–335.

[68] SOUČEK, J. – *Singular solutions to linear elliptic systems,* Comment. Math. Uni. Carolinae **23** (1984), 273–281. preprint.

[69] STAMPACCHIA, G. – *The spaces* $L^{p,\lambda}$, $N^{(p,\lambda)}$ *and interpolation,* Ann. Sc. Norm. Sup. Pisa **19** (1965), 443–462.

[70] STAMPACCHIA, G. – *Equations elliptiques du second ordre à coefficients discontinues,* Les Presses de l'Univ. de Montréal (1966).

[71] STEIN, E.M. – *Singular integrals and differentiability properties of functions,* Princeton Univ. Press, Princeton (1970).

[72] JOST, J., MEIER, M. – *Boundary regularity for minima of certain quadratic functionals,* Math. Ann. **262** (1983), 549–561.

[73] GIAQUINTA, M. – *A counterexample to the boundary regularity of solutions to elliptic quasilinear systems,* manuscripta math. **14** (1978), 217–220.

Index

Lectures in Mathematics - ETH Zürich

Each year the Eidgenössische Technische Hochschule (ETH) at Zürich invites a selected group of mathematicians to give postgraduate seminars in various areas of pure and applied mathematics. These seminars are directed to an audience of many levels and backgrounds. Now some of the most successful lectures are being published for a wider audience through the ***Lectures in Mathematics - ETH Zürich*** *series. Lively and informal in style, moderate in size and price, these books will appeal to professionals and students alike, bringing a quick understanding of some important areas of current research.*

Previously published:

Randall J. LeVeque, Numerical Methods for Conservation Laws.
Second edition 1992, 214 pages, softcover, ISBN 3-7643-2723-5.

J. Donald Monk, Cardinal Functions on Boolean Algebras.
1990, 152 pages, softcover, ISBN 3-7643-2495-3.

Carl de Boor, Splinefunktionen (german).
1990, 184 pages, softcover, ISBN 3-7643-2514-3.

Daniel Bättig/Horst Knörrer, Singularitäten (german).
1991, 140 pages, softcover, ISBN 3-7643-2616-6.

Anthony J. Tromba, Teichmüller Theory in Riemannian Geometry.
1992, 224 pages, softcover, ISBN 3-7643-2735-9.

Raghavan Narasimhan, Compact Riemann Surfaces.
1992, 128 pages, softcover, ISBN 3-7643-2742-1.

Marc Yor, Some Aspects of Brownian Motion. Part I: Some Special Functionals.
1992, 144 pages, softcover, ISBN 3-7643-2807-X.

Olavi Nevanlinna, Convergence of Iterations for Linear Equations.
1993, 184 pages, softcover, ISBN 3-7643-2865-7.

Gilbert Baumslag, Topics in Combinatorial Group Theory.
1993, 172 pages, softcover, ISBN 3-7643-2921-1.

Mariano Giaquinta, Introduction to Regularity Theory for Nonlinear Elliptic Systems.
1993, 144 pages, softcover, ISBN 3-7643-2879-7.

We state some *open problems:*

(a) If possible, improve the estimate of the singular set in theorem 6.6.

(b) What are the properties of the singular set? How is the behaviour of u in a neighbourhood of the singular set?

(c) What can be said about the stability of the singular set under small perturbations?
The set $\{A(x,u) \mid \text{minimizers of } \int_\Omega A(x,u)DuDudx \ \text{ are regular}\}$ is open!

(d) Give reasonable conditions in order that minimizers (or solutions of elliptic quasilinear systems) are everywhere regular.

6.4 The singular set of minimizers of a special class of quadratic functionals

In this final section we consider quadratic functionals of the type

$$\int_\Omega A_{ij}^{\alpha\beta}(x,u)D_\alpha u^i D_\beta u^j dx$$

where the $A_{ij}^{\alpha\beta}$ satisfy the assumptions of the previous section (see p. 113).

Our additional main assumption will now be that the $A_{ij}^{\alpha\beta}$ split, i.e.

$$A_{ij}^{\alpha\beta}(x,u) = g_{ij}(u)G^{\alpha\beta}(x)$$

where (g_{ij}), $(G^{\alpha\beta})$ are symmetric, positive definite matrices and thus give rise to two Riemannian manifolds $(\mathbb{R}^n, (G_{\alpha\beta}))$ and $(\mathbb{R}^N, (g_{ij}))$, $(G_{\alpha\beta}) = (G^{\alpha\beta})^{-1}$. Observe that the functional

$$\int_\Omega G^{\alpha\beta}(x)g_{ij}(u)D_\alpha u^i D_\beta u^j dx$$

has in general no geometric meaning, whereas

$$\mathbb{E}(u) = \int_\Omega G^{\alpha\beta}(x)g_{ij}(u)D_\alpha u^i D_\beta u^j \sqrt{G}dx \qquad \text{for } G = \det(G_{\alpha\beta})$$

expresses the energy for a map $u : (\mathbb{R}^n, (G^{\alpha\beta})) \to (\mathbb{R}^N, (g_{ij}))$.

In the class of such mappings we look for *regular* minima with prescribed boundary condition of the functional $\mathbb{E}(u)$. These are called *harmonic mappings* between the above Riemannian manifolds.

Birkhäuser Advanced Texts - Basler Lehrbücher

Managing Editors:

H. Amann (Universität Zürich)
H. Kraft (Universität Basel)

This series presents, at an advanced level, introductions to some of the fields of current interest in mathematics. Starting with basic concepts, fundamental results and techniques are covered, and important applications and new developments discussed. The textbooks are suitable as an introduction for students and non-specialists, and they can also be used as background material for advanced courses and seminars.

We encourage preparation of manuscripts in TeX for delivery in camera-ready copy which leads to rapid publication, or in electronic form for interfacing with laser printers or typesetters. Proposals should be sent directly to the editors or to: Birkhäuser Verlag, P.O. Box 133, CH-4010 Basel, Switzerland.

Published in the series **Birkhäuser Advanced Texts** - Basler Lehrbücher:

Volume 1: **Markus Brodmann, Algebraische Geometrie** (german).
1989, 470 pages, hardcover, ISBN 3-7643-1779-5.

Volume 2: **Ernest B. Vinberg, Linear Representations of Groups.**
1989, 146 pages, hardcover, ISBN 3-7643-2288-8.

Volume 3: **Konrad Jacobs, Discrete Stochastics.**
1992, 283 pages, hardcover, ISBN 3-7643-2591-7.

Volume 4: **Steven G. Krantz/Harold R. Parks, A Primer of Real Analytic Functions.**
1992, 208 pages, hardcover, ISBN 3-7643-2768-5.

Volume 5: **Lawrence Conlon, Differentiable Manifolds; A First Course**
1993, 369 pages, hardcover, ISBN 3-7643-3626-9.

Monographs in Mathematics

The foundations of this outstanding book series were laid in 1944. Until the end of the 1970s, a total of 77 volumes appeared, including works of such distinguished mathematicians as Carathéodory, Nevanlinna, and Shafarevich, to name a few. The series came to its present name and appearance in the 1980s. According to its well-established tradition, only monographs of excellent quality will be published in this collection. Comprehensive, in-depth treatments of areas of current interest are presented to a readership ranging from graduate students to professional mathematicians. Concrete examples and applications both within and beyond the immediate domain of mathematics illustrate the import and consequences of the theory under discussion.

We encourage preparation of manuscripts in TeX for delivery in camera-ready copy which leads to rapid publication, or in electronic form for interfacing with laser printers or typesetters. Proposals should be sent directly to the editors or to: Birkhäuser Verlag, P.O. Box 133, CH-4010 Basel, Switzerland.

Published in the series Monographs in Mathematics since 1983

Volume 78: **Hans Triebel, Theory of Function Spaces.**
1983, 284 pages, hardcover, ISBN 3-7643-1381-1.

Volume 79: **Gennadi M. Henkin/Jürgen Leiterer, Theory of Functions on Complex Manifolds.**
1984, 228 pages, hardcover, ISBN 3-7643-1477-X.

Volume 80: **Enrico Giusti, Minimal Surfaces and Functions of Bounded Variation.** 1984, 240 pages, hardcover, ISBN 3-7643-3153-4.

Volume 81: **Robert J. Zimmer, Ergodic Theory.**
1984, 210 pages, hardcover, ISBN 3-7643-3184-4.

Volume 82: **V. I. Arnold/S. M. Gusein-Zade/A. N. Varchenko, Singularities of Differentiable Maps - Volume I.**
1985, 392 pages, hardcover, ISBN 3-7643-3187-9.

Volume 83: **V. I. Arnold/S. M. Gusein-Zade/A. N. Varchenko, Singularities of Differentiable Maps - Volume II.**
1988, 500 pages, hardcover, ISBN 3-7643-3185-2.

Volume 84: **Hans Triebel, Theory of Function Spaces II.**
1992, 380 pages, hardcover, ISBN 3-7643-2639-5.

Volume 85: **K.R. Parthasarathy, An Introduction to Quantum Stochastic Calculus.**
1992, 300 pages, hardcover, ISBN 3-7643-2697-2.

Volume 86: **Masao Nagasawa, Schrödinger Equations and Diffusion Theory.**
1993, 332 pages, hardcover, ISBN 3-7643-2875-4.

Volume 87: **Jan Prüss, Evolutionary Integral Equations and Applications.**
1993, 392 pages, hardcover, ISBN 3-7643-2876-2.